Denny Thiemig

Electrocodeposition of Metal Matrix Nanocomposites

Denny Thiemig

Electrocodeposition of Metal Matrix Nanocomposites

Investigation on the Mechanism of Electrocodeposition and the Structure-Properties Correlation of Nickel Nanocomposites

Südwestdeutscher Verlag für Hochschulschriften

Impressum/Imprint (nur für Deutschland/ only for Germany)
Bibliografische Information der Deutschen Nationalbibliothek: Die Deutsche Nationalbibliothek
verzeichnet diese Publikation in der Deutschen Nationalbibliografie; detaillierte bibliografische
Daten sind im Internet über http://dnb.d-nb.de abrufbar.
Alle in diesem Buch genannten Marken und Produktnamen unterliegen warenzeichen-, marken-
oder patentrechtlichem Schutz bzw. sind Warenzeichen oder eingetragene Warenzeichen der
jeweiligen Inhaber. Die Wiedergabe von Marken, Produktnamen, Gebrauchsnamen,
Handelsnamen, Warenbezeichnungen u.s.w. in diesem Werk berechtigt auch ohne besondere
Kennzeichnung nicht zu der Annahme, dass solche Namen im Sinne der Warenzeichen- und
Markenschutzgesetzgebung als frei zu betrachten wären und daher von jedermann benutzt
werden dürften.

Verlag: Südwestdeutscher Verlag für Hochschulschriften Aktiengesellschaft & Co. KG
Dudweiler Landstr. 99, 66123 Saarbrücken, Deutschland
Telefon +49 681 37 20 271-1, Telefax +49 681 37 20 271-0, Email: info@svh-verlag.de
Zugl.: Dresden, TU Dresden, Diss., 2009

Herstellung in Deutschland:
Schaltungsdienst Lange o.H.G., Berlin
Books on Demand GmbH, Norderstedt
Reha GmbH, Saarbrücken
Amazon Distribution GmbH, Leipzig
ISBN: 978-3-8381-0522-2

Imprint (only for USA, GB)
Bibliographic information published by the Deutsche Nationalbibliothek: The Deutsche
Nationalbibliothek lists this publication in the Deutsche Nationalbibliografie; detailed
bibliographic data are available in the Internet at http://dnb.d-nb.de.
Any brand names and product names mentioned in this book are subject to trademark, brand or
patent protection and are trademarks or registered trademarks of their respective holders. The
use of brand names, product names, common names, trade names, product descriptions etc.
even without a particular marking in this works is in no way to be construed to mean that such
names may be regarded as unrestricted in respect of trademark and brand protection legislation
and could thus be used by anyone.

Publisher:
Südwestdeutscher Verlag für Hochschulschriften Aktiengesellschaft & Co. KG
Dudweiler Landstr. 99, 66123 Saarbrücken, Germany
Phone +49 681 37 20 271-1, Fax +49 681 37 20 271-0, Email: info@svh-verlag.de

Copyright © 2009 by the author and Südwestdeutscher Verlag für Hochschulschriften
Aktiengesellschaft & Co. KG and licensors
All rights reserved. Saarbrücken 2009

Printed in the U.S.A.
Printed in the U.K. by (see last page)
ISBN: 978-3-8381-0522-2

TABLE OF CONTENTS I
LIST OF ORIGINAL PUBLICATIONS IV
LIST OF FIGURES VI
LIST OF TABLES XI
LIST OF ABBREVIATIONS AND SYMBOLS XII

0 Introduction and Motivation 1
1 Theoretical Background 4
1.1 Electrocodeposition of Composites 4
1.2 Colloidal Dispersions 6
1.2.1 Electrical Double Layer 6
1.2.2 Dispersion Stability 8
1.3 Electrocodeposition Mechanisms and Models 10
1.3.1 Model of Guglielmi 10
1.3.2 Model of Celis et al. 11
1.3.3 Trajectory Model of Fransaer et al. 13
1.3.4 Kinetic Model of Vereecken et al. 14
1.3.5 Model of Lee and Talbot 15
1.3.6 Summary 15
1.4 Process Variables 16
1.4.1 Electrochemical Cell Configuration and Electrode Orientation 16
1.4.2 Hydrodynamics, Magneto-hydrodynamics 17
1.4.3 Bath Properties 18
1.4.3.1 Electrolyte Composition 19
1.4.3.2 Particle Characteristics 19
1.4.4 Current Density and Current Modulation 20
1.5 Structure and Properties of Metal Matrix Nanocomposites 21
2 Experimental 22
2.1 Deposition Procedures 22
2.1.1 Electrolyte Composition and Working Conditions 22
2.1.2 Nanoparticles 23
2.1.2.1 Commercial Ceramic Nanoparticles 23
2.1.2.2 Synthesis of Magnetic Nanoparticles 24
2.1.3 Experimental Set-up 25
2.1.3.1 Parallel Plate Electrode (PPE) 25
2.1.3.2 Impinging Jet Electrode (IJE) 26
2.1.3.3 Electrocodeposition in a Magnetic Field 28
2.1.3.4 Electrochemical Quartz Crystal Microbalance (EQCM) 29
2.1.4 Substrate Preparation 30

- 2.2 Particle-Characterization 31
  - 2.2.1 Zeta Potential Measurement 31
  - 2.2.2 Photon Correlation Spectroscopy 32
  - 2.2.3 Dispersion Stability 34
- 2.3 Layer-Characterization 35
  - 2.3.1 Particle Incorporation Analysis 35
    - 2.3.1.1 Electrogravimetric Analysis 35
    - 2.3.1.2 Scanning Electron Microscopy (SEM) and energy-dispersive X-ray Spectroscopy (EDX) 36
    - 2.3.1.3 Glow Discharge Optical Emission Spectrometry (GD-OES) 37
  - 2.3.2 Transmission Electron Microscopy (TEM) 38
  - 2.3.3 X-ray Diffraction (XRD) 38
  - 2.3.4 Vickers Microhardness 39
  - 2.3.5 Abrasion Resistance 39
  - 2.3.6 Magnetization Measurements 40

3 Results and Discussion 41
- 3.1 Particle Characterization 41
  - 3.1.1 Alumina 41
  - 3.1.2 Titania 42
  - 3.1.3 Cobalt nanoparticles 45
  - 3.1.4 Magnetite nanoparticles 46
- 3.2 Alumina Particle Adsorption on the Nickel Electrode 48
- 3.3 Electrocodeposition of Ni-Al_2O_3 and Ni-TiO_2 with the Parallel Plate Electrode 51
  - 3.3.1 Direct Current Deposition 51
  - 3.3.2 Pulse Plating and Pulse Reverse Plating 53
- 3.4 Electrocodeposition of Ni-Al_2O_3 with the Impinging Jet Electrode 57
- 3.5 Electrocodeposition of Nickel Matrix Nanocomposites in a Magnetic Field 62
- 3.6 Structure and Properties 65
  - 3.6.1 Surface Morphology 65
  - 3.6.2 Microstructure 67
  - 3.6.3 Particle Incorporation Behavior 71
  - 3.6.4 Mechanical Properties 79
    - 3.6.4.1 Vickers Microhardness 79
    - 3.6.4.2 Abrasion Resistance 80
  - 3.6.5 Magnetic Properties 81

4 Model of Electrocodeposition using an Impinging Jet Electrode 84
- 4.1 Mathematical Model 84
- 4.2 Comparison with Experimental Results 91
- 4.3 Summary 94

5 Conclusions and Suggested Future Work XV
    5.1   Conclusions for the Particle Characterization XV
    5.2   Conclusions for the Electrocodeposition of Nickel Composite Films XV
    5.3   Conclusions for the Structure and Properties of Nickel Composite Films XVII
    5.4   Conclusions for the Modeling of the Electrocodeposition XVIII
    5.5   Suggested Future Work XVIII

6 References XX

List of Original Publications

This thesis is based on the following publications and presentations, respectively.

A. Bund, **D. Thiemig**
„Electrodeposition of Cu/alumina and Ni/alumina nanocomposites"
Trans. Electrochem. Soc. **3** (2006) 85-94

A. Bund, **D. Thiemig**
„Influence of bath composition and pH on the electrocodeposition of alumina nanoparticles and nickel"
Surf. Coat. Technol. **201** (2007) 7092-7099

D. Thiemig, A. Bund, J.-B. Talbot
„Electrocodeposition of Nickel Nanocomposites using an Impinging Jet Electrode"
J. Electrochem. Soc. **154** (2007) D510-D515

D. Thiemig, R. Lange, A. Bund
„Influence of pulse plating parameters on the electrocodeposition of metal matrix nanocomposites"
Electrochim. Acta **52** (2007) 7362-7371

D. Thiemig, R. Lange, A. Bund
„Der Einfluss modulierter Ströme auf die Dispersionsabscheidung von Nanokompositen"
Galvanotechnik **106** (2007) 2103-2112

D. Thiemig, A. Bund
„Characterization of electrodeposited Ni-TiO$_2$ nanocomposite coatings"
Surf. Coat. Technol. **202** (2008) 2976-2984

D. Thiemig, A. Bund, J.-B. Talbot
„Influence of Hydrodynamics and Pulse Plating Parameters on the Electrocodeposition of Nickel Nanocomposite Films"
Electrochim. Acta (accepted)

D. Thiemig, C. Kubeil, C. P. Gräf, A. Bund
„Electrocodeposition of Magnetic Nickel Matrix Nanocomposites in a Static Magnetic Field"
Thin Solid Films (submitted)

D. Thiemig, A. Bund, J.-B. Talbot
„Model of electrocodeposition using an unsubmerged impinging jet electrode"
J. Electrochem. Soc. (submitted)

D. Thiemig, A. Bund, J.-B. Talbot
„Untersuchungen zur elektrochemischen Herstellung und zu den Struktur-Eigenschaftsbeziehungen von Nickel Dispersionsschichten"
Galvanotechnik (submitted)

D. Thiemig, R. Lange, A. Bund
„Electrodeposition of Cu/alumina and Ni/alumina nanocomposites"
210th Meeting of the Electrochemcial Society, 29.10.-03.11.2006, Cancun, oral presentation.

D. Thiemig, R. Lange, A. Bund
„Pulse Plating of Metal Matrix Nanocomposites"
58th Annual Meeting of the International Society of Electrochemistry, 9.-14.09.2007, Banff, oral presentation.

D. Thiemig, R. Lange, A. Bund
„Der Einfluss modulierter Ströme auf die Dispersionsabscheidung von Nanokompositen"
DGO Oberflächentage 2007, 19.-21.09.2007, Garmisch Partenkirchen, oral presentation.

D. Thiemig, A. Bund, J.-B. Talbot
„Influence of Hydrodynamics and Pulse Plating Parameters on the Electrocodeposition of Nickel Nanocomposite Films"
EuroInterfinish 2007, 18.-19.10.2007, Athens, oral presentation.

D. Thiemig, A. Bund
„Electrocodeposition and Characterization of Nickel Nanocomposite Films"
IX. Symposium of Colloids and Surface Chemistry, 29.–31.05.2008, Galati, oral presentation.

D. Thiemig, A. Bund, J.-B. Talbot
„Electrocodeposition of Nickel Nanocomposite Films"
Gordon Research Conference - Electrodeposition, 27.07.–01.08.2008, New London, poster presentation.

The publications are reprinted with the permission of the publishers.

List of Figures

Figure 1   Schematic of the parallel plate electrocodeposition process 4

Figure 2   Schematic representation of the multilayer structure and the potential profile of the electrical double layer at the particle-solution interface in an aqueous electrolyte. 7

Figure 3   Depiction of Guglielmi's electrocodeposition model 10

Figure 4   Depiction of the five-step electrocodeposition mechanism 12

Figure 5   Forces acting on a rigid spherical particle in the vicinity of the electrode 13

Figure 6   Experimental setup of the parallel plate electrode system 25

Figure 7   Experimental of the unsubmerged impinging jet electrode system 26

Figure 8   Schematic of the electroplating cell of the IJE system 27

Figure 9   Schematic of the experimental setup of the electrocodeposition in a (a) parallel (x-direction) and (b) perpendicular (z-direction) magnetic field (**B** field). WE working electrode, CE counter electrode. The black sphere represents the particle and j(Ni^{2+}) is the electric current carried by the Ni deposition. f_L and f_{mp} are the Lorentz and magnetophoretic force, respectively 28

Figure 10 Schematic of the impinging jet cell used for the EQCM experiments 29

Figure 11 Schematic representation of the principle of the microelectrophoretic measurement 31

Figure 12 Schematic representation of the principle of the photon correlation spectroscopy 32

Figure 13 Schematic of the measuring principle of the LUMIfuge 114 34

Figure 14 Schematic of the measuring principle of the Taber Linear Abraser 39

Figure 15 High-resolution bright field TEM image of the as-received 13 nm Al_2O_3 particles 41

Figure 16 Zeta potential of 13nm Al_2O_3 nanoparticles (0.2 g l^{-1}) in diluted 10^{-3} M electrolytes as a function of pH. (■) KCl; (●) sulfamate electrolyte; (▲) pyrophosphate electrolyte 42

Figure 17 High-resolution bright field TEM image of the as-received TiO_2 particles 42

Figure 18 Zeta potential of TiO_2 nanoparticles (0.2 g l^{-1}) in diluted 10^{-3} M electrolytes as a function of pH. (■) KCl; (●) sulfamate electrolyte; (▲) pyrophosphate electrolyte 43

List of Figures

Figure 19 Zeta potential of TiO_2 nanoparticles (0.2 g l^{-1}) in different electrolytes as a function of pH (all concentrations are 10^{-3} M). (a): (■) nickel sulfamate; (●) nickel chloride; (▲) boric acid; (b): (■) nickel sulphate; (●) potassium pyrophosphate; (▲) citric acid 44

Figure 20 Sedimentation curves of TiO_2 nanoparticles (5 g l^{-1}) in different electrolytes at a rotation speed of 500 rpm. (■) sulfamate electrolyte; (●) pyrophosphate electrolyte 45

Figure 21 High-resolution bright field TEM image of the as-prepared Co nanoparticles. (a) cubes; (b) discoid 45

Figure 22 High-resolution bright field TEM image of the as-prepared magnetite nanoparticles 46

Figure 23 XRD pattern of Fe_3O_4 nanoparticles. (—) measured pattern; (—) reference pattern (JCPDS no. 19-0629) 47

Figure 24 Hysteresis loop of the as-prepared Fe_3O_4 nanoparticles 47

Figure 25 Zeta potential and hydrodynamic diameter of the Fe_3O_4 nanoparticles in the diluted nickel pyrophosphate electrolyte as a function of pH 48

Figure 26 Mass adsorbed onto a freshly nickel coated quartz as a function of (a) electrolyte flow rate at an electrode potential of -750 mV vs. Ag/AgCl and (b) electrode potential at a flow rate of 600 ml min^{-1}. Electrolyte composition: 5 g l^{-1} Al_2O_3 in 1 M KCl. (□) pH 4.3; (○) pH 7.5; (Δ) pH 9.5 49

Figure 27 Current density transients during the EQCM adsorption measurements at a flow rate of 600 ml min^{-1}, a pH of 4.3 and an electrode potential of (—) 0 or (—) -1000 mV vs. Ag/AgCl. Electrolyte composition: 5 g l^{-1} Al_2O_3 in 1 M KCl 50

Figure 28 Correlation between the Al_2O_3 content in the layer and the particle loading of 13 nm Al_2O_3 particles. (□) 1; (○) 5; (Δ) 10 A dm^{-2} 51

Figure 29 Correlation between the TiO_2 content in the layer and the concentration of TiO_2 in the electrolyte for different current densities. (■) 1; (●) 5; (▲) 10 A dm^{-2}: (a) sulfamate electrolyte; (b) pyrophosphate electrolyte 52

Figure 30 Effect of duty cycle (a) and pulse frequency (b) on the alumina content of Ni-Al_2O_3 composites. Particle loading: (■) 1; (●) 5; (▲) 10 g l^{-1} of 13 nm Al_2O_3 particles. The range of incorporation data using DC deposition at various current densities and particle loadings is indicated by the hatched areas 54

Figure 31 Effect of the average current density during PP on the particle incorporation in nickel for different particle loadings. (□) 1; (○) 5; (Δ) 10 g l^{-1} of 13 nm Al_2O_3 particles 55

List of Figures

Figure 32   Effect of cathodic pulse time during PRP on the particle content of Ni-Al_2O_3 composites prepared at t_{rev}=20 ms from electrolytes containing 10 g l^{-1} of 13 nm Al_2O_3 particles. i_p=5 A dm^{-2}: (■) i_{an}=1 A dm^{-2}; (●) i_{an}=5 A dm^{-2}; i_p=10 A dm^{-2}: (□) i_{an}=1 A dm^{-2}; (○) i_{an}=5 A dm^{-2}. The range of incorporation data using PPE configuration at various current densities and particle loadings of 13 nm Al_2O_3 particles is indicated by the hatched area 56

Figure 33   Correlation between the Al_2O_3 content in the layer and IJE flow rate for different particle loadings of the electrolyte at a current density of 10 A dm^{-2}. (■) 90, (●) 120 g l^{-1} 50 nm Al_2O_3 particles 58

Figure 34   Effect of particle loading of the electrolyte on the particle incorporation at a current density of 10 A dm^{-2}. (■) 1; (●) 2.5; (▲) 6.5 l min^{-1} 58

Figure 35   Influence of current density on the incorporation of alumina particles in nickel at particle loading of 90 g l^{-1} Al_2O_3. (■) 1; (●) 2.5; (▲) 6.5 l min^{-1} 59

Figure 36   Particle incorporation vs. percentage of limiting current density at a particle loading of 90 g l^{-1} of 50 nm Al_2O_3 particles. (□) 10; (○) 15; (Δ) 20 A dm^{-2} .. 59

Figure 37   Particle content of the Ni-Al_2O_3 nanocomposites determined by EDX analysis as a function of the incorporation results of the electrogravimetric method.. 60

Figure 38 Al_2O_3 content (determined by means of EDX analysis) in a nickel film produced via Jet-Plating at a current density of 10 A dm^{-2} from an acidic sulfamate electrolyte containing 90 g l^{-1} 50 nm Al_2O_3 nanoparticles. (■) 1; (●) 2.5; (▲) 6.5 l min^{-1}. The inside diagrams indicate the depth profile of the nanoparticle distribution in the film obtained by GD-OES analyses 61

Figure 39   Comparison of the particle content of Ni-Fe_3O_4 composites plated at different current densities from electrolytes containing 10 g l^{-1} cubic cobalt nanoparticles. (□) ***B*** = 0 mT; (○) ***B*** = 100 mT, perpendicular to i (x-direction); (Δ) ***B*** = 100 mT, parallel to i (z-direction, see Fig. 9b, p. 28) 63

Figure 40   SEM surface morphology of pure Ni films plated at 5 A dm^{-2}; (a) sulfamate bath; (b) pyrophosphate bath 65

Figure 41   Surface morphology of Ni and Ni-Al_2O_3 composites plated at θ = 67 %, f = 1.67 Hz. 0 g l^{-1} Al_2O_3: (a) 10 A dm^{-2}, (c) 5 A dm^{-2}. 10 g l^{-1} Al_2O_3: (b) 10 A dm^{-2}, (d) 5 A dm^{-2} 66

Figure 42   Cross-sectional SEM images of pure nickel films. (a) DC i =10 A dm^{-2}, (b) PP θ = 67 %, f = 3.3 Hz, i_p = 10 A dm^{-2}, (c) IJE 6.5 l min^{-1}, i =10 A dm^{-2} 67

List of Figures

Figure 43 Cross-sectional SEM images of a Ni-Al$_2$O$_3$ composite film plated at 10 A dm^{-2}, 6.5 l min^{-1} and 60 g l^{-1} of 50 nm Al$_2$O$_3$ particles in the electrolyte. (with increasing resolution) 68

Figure 44 Cross sectional SEM image of pure nickel (a) and a Ni-TiO$_2$ nanocomposite (b) film deposited from pyrophosphate bath. (a) 10 A dm^{-2}; (b) 5 A dm^{-2}, 10 g l^{-1} TiO$_2$ in the electrolyte 68

Figure 45 X-ray diffraction patterns of Ni and Ni-TiO$_2$ films plated at a current density of 1 A dm^{-2}. (a) acidic sulfamate bath, (b) alkaline pyrophosphate bath 69

Figure 46 QBSD image (a) and IPF color coded nickel map (b) of a pure nickel film deposited from the sulfamate electrolyte by PRP: i$_P$=5 A dm^{-2}, t$_{on}$=400 ms and i$_{an}$=1 A dm^{-2}, t$_{rev}$=20 ms 73

Figure 47 QBSD image (a) and IPF color coded nickel map (b) as well as detail in image quality map (c) IPF color coded nickel map (d) and IPF color coded alumina map (e) of a Ni-Al$_2$O$_3$ film deposited by PRP (same plating conditions as Fig. 46, 10 g l^{-1} 13 nm Al$_2$O$_3$ in the electrolyte). 73

Figure 48 QBSD image (a) and IPF color coded nickel map (b) of a Ni-Al$_2$O$_3$ film deposited from the sulfamate electrolyte containing 10 g l^{-1} 13 nm Al$_2$O$_3$ by DC: i=10 A dm^{-2} 74

Figure 49 Pole figures of composite films plated from electrolyte containing 10 g l^{-1} 13 nm Al$_2$O$_3$. (a) DC: 10 A dm^{-2}; (b) PRP: i$_P$=5 A dm^{-2}, t$_{on}$=400 ms and i$_{an}$=1 A dm^{-2}, t$_{rev}$=20 ms 74

Figure 50 SE (a, b) and STEM (c, d) dark field images of a Ni-Al$_2$O$_3$ film deposited by PRP (same plating conditions as Fig. 46, 10 g l^{-1} 13 nm Al$_2$O$_3$ in the electrolyte) 76

Figure 51 TEM image of a Ni-Al$_2$O$_3$ film produced by PRP (same plating conditions as Fig. 46, 10 g l^{-1} 13 nm Al$_2$O$_3$ in the electrolyte) 77

Figure 52 High resolution TEM images of a Ni-Al$_2$O$_3$ film produced by PRP (same plating conditions as Fig. 46, 10 g l^{-1} 13 nm Al$_2$O$_3$ in the electrolyte) 77

Figure 53 High resolution TEM image of a Ni-Al$_2$O$_3$ film (same plating conditions as Fig. 46, 10 g l^{-1} 13 nm Al$_2$O$_3$ in the electrolyte) including the subscription of certain lattice planes of the nickel matrix 78

Figure 54 Correlation between the Vickers microhardness and the TiO$_2$ content of the layer; (■) 1 A dm^{-2}; (●) 5 A dm^{-2}; (▲) 10 A dm^{-2}. (a) sulfamate electrolyte; (b) pyrophosphate electrolyte 79

List of Figures

Figure 55 Weight loss after bi-directional sliding test as a function of the plating current density. Sulfamate electrolyte: (■) 0 g l^{-1}; (●) 10 g l^{-1} TiO$_2$; Pyrophosphate electrolyte (□) 0 g l^{-1}; (○) 10 g l^{-1} TiO$_2$ 81

Figure 56 Hysteresis loops of pure Ni films deposited at 5 A dm^{-2}. (■) B = 0 mT; (●) B = 100 mT (z-direction); (▲) B = 100 mT, (x-direction, Fig. 9, p. 28) 82

Figure 57 Schematic representation of the particle concentration in the impinging region of the unsubmerged IJE system. $c_{p,b}$, $c_{i,b}$ are the bulk concentration and $c_{p,s}$, $c_{i,s}$ are the concentration at the electrode surface of the particle and reactive ion, respectively 84

Figure 58 Particle diffusion boundary layer thickness as a function of the electrolyte velocity 88

Figure 59 Parameter A_D (IJE flow rate: — 1 l min^{-1}, — 2.5 l min^{-1}, — 6.5 l min^{-1}) and A_G (—) in Eq. 24 vs. particle diameter 89

Figure 60 The number of nanoparticles per unit volume of nickel in the film, m_p, vs. particle number density in the solution, n_b, for flow rates of 1-6.5 l min^{-1}. The solid lines are the calculated particle number densities in the film under diffusion-limiting conditions (IJE flow rate: (■) 1 l min^{-1}, (●) 2.5 l min^{-1}, (▲) 6.5 l min^{-1}) and the open squares refer to the experimental results. (a) 5 A dm^{-2}, (b) 10 A dm^{-2}, (c) 15 A dm^{-2}, (d) 20 A dm^{-2} 91

Figure 61 Particle size distribution of the alumina nanoparticles in the acidic nickel sulfamate electrolyte determined by photon correlation spectroscopy 92

Figure 62 The number of nanoparticles per unit volume of nickel in the film, m_p, vs. particle number density in the solution, n_b, for IJE flow rates of 1-6.5 l min^{-1} and a current density of (a) 5 A dm^{-2} or (b) 20 A dm^{-2}, respectively. The hatched areas refer to the theoretical particle number densities in the film taking into account the actual particle size distribution in the electrolyte (Fig. 61) and the open squares refer to the experimental results 93

List of Tables

Table 1	Electrolyte composition and working conditions	22
Table 2	Particle properties as given by the supplier	23
Table 3	Applied pulse plating current programs. The cathodic peak current density was 5 and 10 A dm^{-2}, respectively	26
Table 4	Particle content of the Ni-Co composites prepared from electrolytes containing 6 g l^{-1} Co particles of either disk-like or cubically shape. **B** refers to the direction of the 100 mT magnetic field during electrodeposition: 0 no field; x - perpendicular to i; z - parallel to i (see Fig. 9b, p. 28)	62
Table 5	Relative texture coefficient RTC$_{hkl}$ of the pure nickel and Ni-TiO$_2$ composite films deposited from electrolytes containing 10 g l^{-1} TiO$_2$	70
Table 6	Crystallite size of pure nickel and Ni-TiO$_2$ composites, deposited from electrolytes containing 10 g l^{-1} nanoparticles, determined by the Scherrer line broadening using the (200) reflection of the XRD pattern	70
Table 7	Grain features of the nickel matrix	72
Table 8	Magnetic properties of pure Ni and Ni-Co composites. Composites have been prepared from electrolytes containing 6 g l^{-1} Co particles	82
Table 9	List of values for physical properties	88

List of Abbreviations

CE	counter electrode
DC	direct current
EBSD	electron backscattered diffraction
ECD	electrocodeposition
EDX	energy dispersive X-ray spectroscopy
EDL	electrical double layer
EQCM	electrochemical quartz crystal microbalance
EsB	elastically scattered back scattered electrons
FHWM	full width at half maximum
GD-OES	glow discharge optical emission spectrometry
IJE	impinging jet electrode
MEMS	micro-electromechanical systems
MHD	magnetohydrodynamic effect
MMNC	metal matrix nanocomposite
OP	oxide polish
PCS	photon correlation spectroscopy
PP	pulse plating
PPE	parallel plate electrode
PRP	pulse reverse plating
QBSD	quadrant backscattered diffraction
RCE	rotating cylinder electrode
RDE	rotating disk electrode
RE	reference electrode
RTC	relative texture coefficient
SCE	saturated calomel electrode
SE	secondary electron
SEM	scanning electron microscopy
STEM	scanning transmission electron microscopy
TEM	transmission electron microscopy
VSM	vibrating sample magnetometry
WE	working electrode
XRD	X-ray diffraction

List of Symbols

A_D	transport parameter for diffusion flux
A_G	transport parameter for gravitational flux corrected for the buoyancy
B	magnetic field, T
$c_{i,b}$	concentration of ionic species in the bulk electrolyte solution, mol m^{-3}

List of Abbreviations and Symbols

$c_{i,s}$	concentration of ionic species at the electrode surface, mol m^{-3}
$c_{p,b}$	concentration of particles in the bulk solution, mol m^{-3}
$c_{p,s}$	concentration of particles at the electrode surface, mol m^{-3}
d	jet nozzle diameter, m
D_m	diffusion coefficient of the reactive metal species, m^2 s^{-1}
D_p	particle diffusion coefficient, m^2 s^{-1}
f	friction coefficient ($f = 6\pi\eta r$)
f_L	Lorentz force, N
f_{mp}	magnetophoretic force, N
f_P	pulse frequency, s^{-1}
$f_{R,0}$	resonance frequency of the unloaded quartz, Hz
F	Faraday's constant, 9.64853 x 10^4 A s mol^{-1}
F_{Diff}	diffusion force, N
F_{Grav}	gravitation force, N
F_{total}	total force acting on the particle, N
g	gravitation acceleration rate, m^2 s^{-1}
H	nozzle-substrate gap distance, m
H_C	coercivity, Oe
i	current density, A dm^{-2}
i_{ave}	average current density, A dm^{-2}
i_{an}	anodic peak current density, A dm^{-2}
i_P	cathodic peak current density, A dm^{-2}
i_L	limiting current density, A dm^{-2}
J_p	mass flux of particles on an electrode surface, mol m^{-2} s^{-1}
J_m	mass flux of deposited metal atoms on an electrode surface, mol m^{-2} s^{-1}
k	average mass transfer coefficient ($k = D_m / \delta$), m s^{-1}
k_B	Boltzmann's constant, 1.380658 x 10^{-23} J K^{-1}
m	single particle mass, kg
m_p	number of particles per unit volume of nickel in the film
n_b	number density of particles per unit volume of electrolyte solution
n_m	number of moles of metal atoms in a composite film, mol
n_p	number of moles of particles in a composite film, mol
M_S	saturation magnetization, m emu
N_A	Avogadro's number, 6.022136 x 10^{23} mol^{-1}
r	particle radius, m
R	gas constant, J mol^{-1}K^{-1}
Re	Reynolds number
Sc	Schmidt number
Sh	Sherwood number

t_{on}	pulse on-time, s
t_{off}	pulse pause, s
t_{rev}	pulse reverse time, s
T	temperature, K
υ	jet flow velocity, m s^{-1}
υ_{Dr}	net drift velocity of the particles, m s^{-1}
$V_{m,M}$	molar volume of the deposited metal, m^3 mol^{-1}
V_p	single particle volume, m^3
x_v	volume fraction of inert particles incorporated in a composite film, vol-%
z	charge number

Greek letters

δ	boundary layer thickness for metal ions, m
δ_0	viscous sublayer boundary layer thickness, m
δ_p	boundary layer thickness for particles, m
ρ_f	density of the electrolyte solution, kg m^{-3}
ρ_p	density of a particle, kg m^{-3}
∇c	concentration gradient
$\nabla \mu$	chemical potential gradient
κ^{-1}	Debye length, m^{-1}
η	dynamic viscosity of the electrolyte solution, kg m^{-1} s^{-1}
ν	kinematic viscosity of the electrolyte solution, m^{2}s^{-1}
θ	duty cycle, %
ζ	zeta potential, mV

Subscripts

b	bulk electrolyte solution
m	deposited metal species
p	particle
s	electrode surface

0 Introduction and Motivation

Metal matrix nanocomposites (MMNCs) consisting of ultrafine particles of pure metals, ceramics and organic materials in a metal matrix have attracted extensive attention from science and technology since decades [1-3]. Due to their beneficial electrical [4, 5], optical [4, 6, 7], magnetic [8, 9] and mechanical [10-13] properties, they are promising candidates for advanced materials [14, 15]. A variety of preparation techniques, such as thermal and plasma spraying, combination of chemical and physical vapor deposition, powder metallurgy or stir casting have been investigated for developing MMNCs. Among those the process of electrocodeposition (ECD), *i.e.* the process of particle incorporation during the electrolytic deposition of metal, has the advantages of uniform depositions on complexly shaped substrates, low cost, good reproducibility, homogenous distribution of particles, ability of continuous processing, and the reduction of waste [2]. However, there are also several challenges. Dispersal of the particles in a common plating bath can be problematic. In most cases agglomeration and sedimentation of the particles occurs which makes successful codeposition difficult.

A variety of nanosized particles, such as Al_2O_3 [2, 16, 17], C [18-20], SiO_2 [21-24], ZrO_2 [25-27], SiC [28-32] have been successfully co-deposited with several metals. The concentration of particles suspended in the electrolyte has varied from 1 to 200 g l^{-1}, typically producing composites with 1-10 vol% particles [2, 33, 34]. ECD has found use in many important industrial sectors, such as automotive, construction, power generation and aerospace applications [15, 35]. More recently, magnetic micro-electromechanical systems (MEMS) have gained increasing scientific and industrial interests due to their high force and long range application, low power consumption and reversibility [36-42]. A great deal of research effort has been dedicated to the field of soft magnetic materials for sensor implementation [43-45]. However, a higher magnetic moment is required in the case of MEM actuators which are operated by an external magnetic field [37, 46-48]. Therefore, the application of hard magnetic materials seems promising [41, 42, 48]. The application of nanocrystalline magnetic particles for such media seems to be advantageous. Moreover, the incorporation of magnetically hard particles in an electrodeposited Ni or Ni-alloy matrix was found to significantly increase the coercivity of the resulting films [41].

The amount and distribution of the incorporated particles depend upon a variety of inter-related process parameters, including particle characteristics, bath composition, electrochemical cell configuration and operating conditions, particularly hydrodynamics, current density and current modulation [2, 49-54]. Several theoretical models have been proposed to describe the ECD phenomena [1, 49, 51, 55, 56]. However, up to date the mechanism of particle incorporation is not fully understood. This lack of fundamental understanding of the ECD process has resulted in a trial and error approach being the only method of developing parameters for industrial applicable codeposition of composite films with fairly reproducible particle content. Our group has addressed this issue by systematically studying the incorporation behavior of alumina and diamond nanoparticles within a copper, gold and nickel matrix [16, 17, 57]. It has been shown that the surface charge of the nanoparticles is an important parameter for the ECD process. An interesting result was that negatively charged particles are more easily co-deposited compared to positively charged ones. This finding, which is a little bit counterintuitive at the first glance, was explained by an electrostatic model which takes into consideration the presence of ionic clouds of particle and electrode [16, 17, 57].

In the ECD process, bath agitation is usually necessary to maintain a dispersed suspension and to transport the particles to the cathode surface. Most studies on ECD have focused on a parallel plate electrode configuration because of its simplicity [2, 50]. However, due to the various ways of agitating the suspension, an analysis and a comparison of hydrodynamics is almost impossible with this configuration. The control of hydrodynamics can be achieved by using a rotating cylinder (RCE) [58-60] or rotating disk electrode (RDE) [21, 32, 60]. However, both the RCE and the RDE configurations are not typically a viable industrial method, because of their limitation of specific cathode shapes. Another promising way to control hydrodynamics is the impinging jet electrode (IJE) configuration [61]. The IJE provides the advantages of selective [62] and high-speed plating [63]. Furthermore, it is an attractive method to electrodeposit gradient coatings while controlling the volume fraction of particles by changing the jet velocity [64]. There is little research on the ECD of nickel based composites plated with an IJE [64-66]. Recently, ECD with an unsubmerged IJE resulted in particle incorporations of up to 30 vol-% SiC in a nickel matrix [65]. However, no systematic investigation of the effects of the IJE plating parameters on the ECD of Al_2O_3 nanoparticles and nickel has been reported so far.

The properties of MMNCs mainly depend on their composition and structure [11]. Uniform distribution and high amounts of co-deposited particles within the metal matrix were found to be crucial to improve the coating properties [16, 67]. In many cases the enhanced performance of the coatings is mainly caused by a change in the microstructure of the metal matrix [16, 30] and not so much by the presence of the particles themselves. The structure of electroplated metal and alloy coatings is specified by the electrocrystallization process, particularly by the interplay between nucleation and crystal growth [68]. An attractive way of controlling these two processes is the application of a periodically changing current [69]. Compared to direct current (DC) plating, pulse plating (PP) can yield nanocrystalline coatings with improved surface appearance and properties, such as smoothness, refined grains and enhanced corrosion resistance [13, 30, 70-73].

Besides the electrocrystallization of the metal matrix, the particle inclusion during MMNCs plating is affected by pulse plating. It has been observed that PP [53, 54, 74-76] as well as pulse reverse plating (PRP) [26, 77] could be used to enhance the particle incorporation. Furthermore, it has been shown in Refs. [78-80] that PP and PRP can be used for a size-selective incorporation of alumina nanoparticles into copper and nickel matrices, respectively. There are some reports on the effects of PP and PRP on the ECD of alumina with nickel from acidic electrolytes [54, 74, 79-81]. Promising improvements in terms of the amount and the distribution of incorporated alumina particles have been reported with a pulse length in the order of milliseconds [54, 79, 80]. However, no extensive investigations on the effect of the several interrelated PP and PRP parameters can be found in the literature.

The overall objective of this work was to investigate systematically the ECD of nickel nanocomposites using two kinds of nickel plating baths; an acidic sulfamate and an alkaline pyrophosphate bath. Nickel was chosen as a sample matrix material, because it is one of the common industrial coatings used for decorative and functional applications. The electrodeposition of nickel from sulfamate electrolytes is an industrially important process [82]. The advantage over other electrolytes, *e.g.* the Watts bath, include high current efficiency, low film stress even without the addition of additives and the generally good mechanical properties of the nickel deposit [83]. Particle reinforcements have included alumina (Al_2O_3) and titania (TiO_2) since these nanoparticles are readily industrially available in large quantities with almost constant properties. Hence, an up-scaling of the ECD process would be feasible. Moreover, Ni-TiO_2 composites have been reported to be advantageous for high temperature [84] and photoelectrochemical applications [7]. Additionally, the ECD of cobalt (Co) and magnetite (Fe_3O_4) nanoparticles with nickel was studied in the presence of a static magnetic field. The motivation for the application of a magnetic field during deposition was to enhance the particle content of the composite films. Apart from the well established parallel plate direct current deposition, nanocomposites were produced by means of pulse-plating and jet-plating. The effects of the deposition parameters (electrolyte composition, type of current, current density, pH, hydrodynamics, orientation of magnetic field, etc.) on the codeposition of nanoparticles with nickel were investigated by evaluating the particle content, structure as well as the mechanical, *i.e.* microhardness and abrasion resistance, and magnetic properties, *i.e.* saturation magnetization and coercivity, of the composite films.

1 Theoretical Background

The aim of this chapter is to describe the state of knowledge of the ECD mechanism. The first section gives a brief introduction of the ECD process. The subsequent section contains information about the characterization of the particle-electrolyte interface and the forces acting between particles affecting the dispersion stability. The sections 1.3 and 1.4 are dedicated to the most important models that have been proposed so far to describe the ECD process and to certain process parameters affecting the particle incorporation. The last section contains information about the structure of metal matrix nanocomposite films with special regard to materials properties.

1.1 Electrocodeposition of Composites

Metal matrix nanocomposite films (MMNCs) can be obtained by particle incorporation during metal deposition from electrolytes containing a dispersion of ultrafine particles [85]. The electrocodeposition (ECD) process is schematically depicted in Fig. 1 for the parallel plate electrode, which is most commonly used due to its simplicity. In this system, the particles are typically suspended in the plating bath by means of mechanical agitation and typically a small fraction is embedded in the growing metal film.

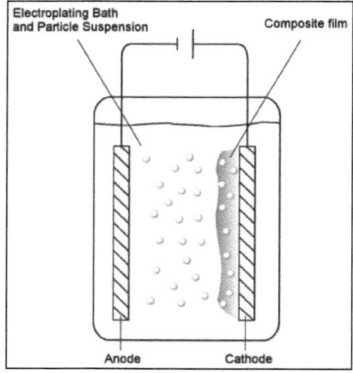

Figure 1 Schematic of the parallel plate electrocodeposition process.

Depending on the combination of matrix and dispersed phase, the coatings may have unique chemical, electrical and mechanical properties [14, 15, 86, 87]. Recently, ECD is industrially used to produce abrasion and corrosion resistant coatings for automotive applications [35]. Moreover, a variety of nanoparticles, such as Al_2O_3 [2, 16, 17], C [18-20], SiO_2 [21-24], ZrO_2 [25-27], SiC [28-32] have been successfully co-deposited with several metals, *e.g.* Cu, Co, Cr, Au, Ni, and Ag, etc. for use in electrocatalysis, wear and corrosion protection, dispersion-strengthening and lubrication of surfaces [34, 35, 88].

Although the origins of ECD lie in the early 1900s [89], most of the applications and modeling developments have been achieved within the last decades. In the early stages, particle incorporation occurred unintentionally without recognizing the possible benefits of the incorporation. As they were considered as impurities degrading the quality of the plated films, efforts were made to prevent their incorporation [2]. As a result, soluble anodes were enclosed in a bag and filters to remove undesirable particles were implemented in the plating cell [2]. Thenceforward the early 1960's, the benefits of the ECD process were established and several industrial applications as coatings for gas turbines used in advanced power generation and aerospace [90], surface coatings for cutting tools with increased strength [91, 92], lubricated surfaces [93], high thermal conductive dispersion-strengthening of materials for actively-cooled structures [94], and high surface area photoelectrodes for electrocatalysis in solar energy conversion [4] have been developed. While most of the ECD research has focused on maximizing the amount of particle incorporation or improving a certain property of the resulting composite film, relatively little effort was dedicated to the fundamental understanding of the ECD mechanism. Nevertheless, from the research to date the influence of a variety of process parameters (current modulation, current density, pH, hydrodynamics, electrolyte composition, etc.) on the ECD process has been determined, typically by evaluating the change in the amount of particle incorporation when that particular parameter is adjusted. Even though the process variables affecting the ECD process have been generally agreed upon, the particular effect of a certain parameter has often been reported contradictory. Furthermore, it is important to note that these process variables are not necessarily independent from one another and that their effect is strongly dependent on the particle-electrolyte combination studied [2, 3]. Detecting the amount and distribution of the particles incorporated in a metal matrix composite is important in understanding the influence that the process parameters have on the deposited film. Several methods have been used for this purpose, including electrogravimetric analysis [95], gravimetric analysis [96], X-ray fluorescence [97], atomic absorption spectrometry [95, 98], infrared reflection absorption spectroscopy [99], microscopic analysis using scanning electron microscopy [100], and energy dispersive X-ray analysis. In general, the literature lacks a discussion concerning the reproducibility and accuracy of the reported incorporation data [2]. Hence, an accurate and reproducible analytical method is needed for verifying the particle incorporation and distribution.

1.2 Colloidal Dispersions

1.2.1 Electrical Double Layer

The surface of any phase is characterized by a separation of positive and negative charge components leading to a region of varying electrical potential [101]. When two phases are placed in contact, in general, a potential difference develops between them [101, 102]. In the case of colloidal dispersions, one of the phases is a continuous medium (usually an electrolyte solution), and the other phase is the colloidal particle that is dispersed in the continuous phase. If a charged colloidal particle is placed in contact with a solution containing both, positive and negative charges, the charges tend to distribute themselves in a non-uniform way at the particle-electrolyte interface resulting in the formation of the electrical double layer (EDL) [101, 102].

The structure of the EDL has been described by various models. The first theory for a quantitative description of the EDL was developed by Helmholtz in 1879 [102]. The Helmholtz model assumes that a layer of counter ions directly binds to the charges in a plane, and their charge exactly compensates that of the surface charges [102]. The electric field generated by the surface charges is accordingly limited to the thickness of this molecular layer of counter ions. Since, such a structure is equivalent to a parallel-plate capacitor; Helmholtz was able to interpret measurements of the double layer capacity. However, although the charge on solids is confined to a surface, the same is not necessarily true of electrolyte solutions, particularly in the case of low electrolyte concentrations [102].

Gouy [103] and Chapman [104] went a step further and considered the thermal motion of the counter ions in the electrolyte. Therefore, this model involves a diffuse layer of charge in the solution. The highest concentration of excess charges is adjacent to the surface, as there is a strong electrostatic force attracting these ions. The ion concentration decreases progressively with increasing distance, since the electric field is partially screened by the ions that are accumulated closer to the surface [102].

In 1924 Stern united the models of Helmholtz and Gouy and Chapman taking into consideration the finite dimension of the ions [102]. Finally, the EDL refers to the electrical potential function and is divided into two layers, one with rapid potential decay near the surface and the other with a more gradual decay farther from the surface [102] as shown in Fig. 2. The inner layer, which is called the compact, Helmholtz, or Stern layer, contains solvent molecules and other species that are specifically absorbed [102]. The locus of the electrical charge centers of these ions is called the inner Helmholtz plane. The solvated ions can only approach the surface at a distance beyond the inner Helmholtz plane. The locus of the electrical centers of the closest solvated ions is called the outer Helmholtz plane. The solvated ions and the surface interact via only long-range electrostatic forces, which are independent of the chemical properties of the ions. The solvated ions are nonspecifically absorbed. Due to thermal agitation, these ions form a region called the diffuse layer [102]. The ion distribution in the diffuse layer was described by the Poisson-Boltzmann equation; a combination of the Poisson equation (electrostatics) and the Boltzmann distribution (statistical mechanics) [103, 104]. Generally, the Poisson equation relates the charge density at a certain place with the electric potential. Due to the fact that the ions in solution are free to move, their distribution, and thus the charge distribution in the liquid, is unknown. As a result the potential cannot be determined by applying solely the Poisson equation. Additionally, the Boltzmann equation, which describes the probability of finding an ion at some particular point in the solution, has to be considered. The extension of the double layer, referred to as the diffuse double layer thickness or the Debye length (κ^{-1}), is inversely proportional to the ion concentration of the electrolyte and to the square of the valency of the ions involved. It is customarily defined as the distance from the particle-electrolyte interface at which the potential has reached the 1/e fraction of its value at the surface [102].

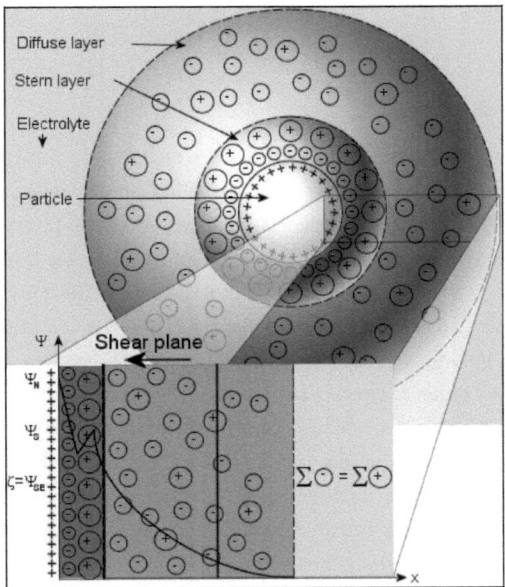

Figure 2   Schematic representation of the multilayer structure and the potential profile of the electrical double layer at the particle-solution interface in an aqueous electrolyte.

The surface of a colloidal particle dispersed in an electrolyte solution is characterized by an EDL consisting of a fixed layer and a diffuse layer as shown in Fig. 2. The interface between the fixed and the diffuse layer is called the shear plane. The zeta potential (ζ) is defined as the potential difference between the shear plane and the bulk solution (Fig. 2) [101]. This potential is taken as an approximation of the surface potential of the particle, since it is not possible to measure the surface potential of the particle itself [101]. The zeta potential is obtained by measuring charged particles in suspension and observing their mobility under an electric field gradient (The method will be discussed in more detail in section 2.2.1.) [101, 105].

The surface charges of colloidal particles can have several origins. The particles might be charged either as a result of imperfections in the crystal structure near the surface, adsorption of specific ions on the surface or dissolution of ionic species from the surface [106]. In the case of metal oxides suspended in an aqueous inert electrolyte, the surface charges are usually generated by the adsorption and desorption, respectively, of protons on hydroxyl groups or hydrogen ions in the surface region. For instance in the case of alumina, the origin for this reaction is based on the incomplete coordination number of the Al^{3+} or the O^{2-} ions in the surface region [101, 106-108].

1.2.2 Dispersion Stability

Dispersion stability refers to the kinetic stability of the particles in suspension. A suspension is considered as stable as long as the individual particles exist separated in solution and do not form aggregates over a relevant time scale [101]. The relevant time scale may vary from tenth of seconds to years, depending on the application [106]. The stability is due to the existence of an energy barrier in the total interaction energy curve preventing the proximity of particles and can be described by the DLVO (**D**erjaguin-**L**andau-**V**erwey-**O**verbeek) theory [109, 110]. As two particles approach, there are at least two types of interaction: the repulsive and the attractive van der Waals interactions, which make most colloids inherently unstable [101]. The van der Waals force is defined as the sum of the Keesom, the Debye and the London dispersion interaction, *i.e.* all the terms that consider dipole interactions. Usually the London dispersion term is dominating [101]. Dispersion stability of particles dispersed in an aqueous solution can be achieved either by electrostatic stabilization (*i.e.* surface charge) or steric stabilization (*i.e.* presence of surfactants or other molecules at the particle surface), or by a combination of both [101, 106]. In the case of an electrostatic stabilized colloid, the repulsion energy barrier depends on the extent of the EDL (1.2.1) and on the magnitude of the surface charge [101]. If the energy barrier is lower for instance at higher ionic strength of the electrolyte or at lower particle charge, the particles may agglomerate irreversibly. An additional important prediction of the DLVO theory is that under certain conditions, a secondary minimum occurs at larger separation distances, which is significant for colloidal systems, since this state of agglomeration (*i.e.* flocculation) is reversible and thus can be easily turned into the dispersed state by *e.g.* mechanical stirring. While the repulsive forces can be tuned via the experimental conditions (*e.g.* pH, ionic strength, etc.), the van der Waals force can just be affected by the type of particle as well as the type of dispersant. If an electrolyte is added to a colloidal suspension, causing marked compression of the electrical double layer around the colloidal particles (section 1.2.1), coagulation will occur. It has been proven by experiments that the DLVO theory is not suitable to describe the stability of all colloidal systems. Moreover, a so-called structural term has to be included which is either attractive or repulsive depending on the particle type [101]. Apart from electrostatic interaction, repulsive interactions can also arise from steric stabilization due to the adsorption of non-ionic polymers onto the particle surface [101].

In an application where particle size is crucial to the process (such as ECD), agglomeration of colloidal particles yielding a higher mean particle size in the suspension can affect the product or process considerably, even if the agglomeration is not enough to cause significant visible settling. In ECD the stability of the colloid dispersion is mandatory, since the particles shall be preferentially incorporated in the form of individual particles in order to improve the film properties [111]. Moreover, the incorporation behavior of particles obviously dependents on their size as will be discussed in section 1.4.3.2. Due to the high ionic strength of the plating electrolytes, the diffuse part of the EDL on the particle will be depressed, leading to a reduced electrostatic repulsion. As a result a lower energy barrier in the total interaction energy curve can cause an irreversible agglomeration of the particles. By measuring fundamental parameters known to affect dispersion stability such as zeta potential (section 2.2.1) and particle size (section 2.2.2), one can evaluate dispersion stability of the sample [101]. The magnitude of the zeta potential gives an indication of the stability of the colloidal system. In general, particles with an absolute value of the zeta potential above 30 mV are usually considered as stable [101].

1.3 Electrocodeposition Mechanisms and Models

The purpose of this chapter is to give a short summary on the main theories which have been developed to describe the ECD process. In general, there are three main steps that must occur during ECD:

(I) Transport of particles from the bulk electrolyte to the electrode surface by means of various mechanisms, *e.g.* convection, diffusion and electrophoresis.
(II) Particle adsorption at the electrode surface.
(III) Irreversible entrapment of the particles in the growing metal layer.

In general, the following discussed models differ in the description of these three stages of the ECD process. A more detailed description of all the models developed for the ECD process can be found in several review papers which describe models, mechanism, effects of process parameters, and possible applications [2, 3, 33-35, 50, 85]. While refined models have been developed during the years, the ECD process is still not understood completely.

1.3.1 Model of Guglielmi

In 1972 Guglielmi proposed the first model of the ECD process which could be experimentally verified [49]. In general, this model is based on two consecutive adsorption steps (Fig. 3).

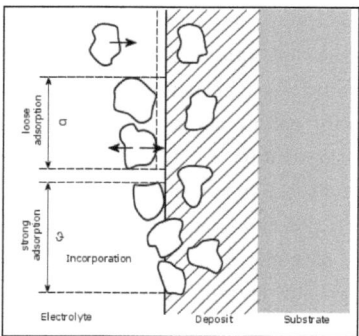

Figure 3 Depiction of Guglielmi's ECD model [49].

During the first step, an initial loose physical adsorption of the particles arriving at the electrode surface takes place. The loosely adsorbed particles are most likely still coated by a thin layer of adsorbed ions and solvent molecules which substantially screens the particle-electrode interactions. The analogy with physical adsorption is reflected by the use of the Langmuir adsorption constant to describe the influence of the particle concentration in the bulk electrolyte on the surface fraction of loosely adsorbed particles.

In the second step which is thought to be field-assisted and electrochemical in nature, the particles become strongly adsorbed onto the electrode due to the applied electrical field. Finally, the strongly adsorbed particles are incorporated in the growing metal matrix.

According to Guglielmi, the volume fraction of incorporated particles, α, is given by

$$\frac{\alpha}{1-\alpha} = \frac{zF\rho_m V_0}{M_m i_0} \cdot e^{(B-A)\eta} \cdot \frac{k\,c_{p,b}}{1+k\,c_{p,b}} \qquad (1)$$

where F is Faraday's constant, ρ_m and M_m are the density and molecular mass of the deposited metal, respectively, $c_{p,b}$ is the particle concentration in the bulk electrolyte, i_0 is the exchange current density, and A is the Tafel constant for the metal reduction reaction. The constants V_0, B and k depend on the particular particle-metal system and have to be determined from ECD experimental data. Guglielmi's model was successfully applied to several ECD systems [49, 112, 113]. However, the model's success is most probably related to the adjustable constants (V_0, B, k), since important process parameters such as hydrodynamics, particle type, size and size distribution, pH and composition of the electrolyte are not explicitly considered in this model.

1.3.2 Model of Celis et al.

Another model was presented by Celis et al. in 1987 who attempted to predict the amount of particle codeposition for a given particle-metal combination by analyzing the particle transfer from the bulk electrolyte to the electrode surface [1]. They proposed a five-step process for the codeposition of particles with metals, which is schematically depicted in Fig. 4. In the first step, the particles in the bulk electrolyte adsorb ions from the electrolyte inducing the formation of an ionic cloud around each particle. In the second step, the particles are transported by means of convection to the hydrodynamic boundary layer. Thence, in the third step, the particles diffuse through the diffusion boundary layer to the electrode surface. Similar to the model of Guglielmi (section 1.3.1) the particles are incorporated in the growing metal film in two consecutive steps. In the fourth step, the particles surrounded by the ionic cloud are adsorbed onto the electrode. The last step is dedicated to the incorporation of the particles in the metal matrix. A successful incorporation can be achieved when a specific quantity, k, of the total quantity of the ions adsorbed on the particle, K, are reduced at the electrode surface. Hence, a certain residence time of a particle at the electrode surface is a precondition for the entrapment. The implementation of a specific residence time is in marked contrast to the basic assumptions of Guglielmi's model which implies that all particles transported to the electrode surface become incorporated.

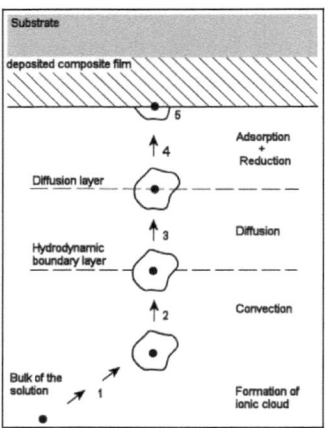

Figure 4 Depiction of the five-step ECD mechanism [1].

Based on this statistical approach, the weight fraction of particle incorporation, $w_{p,th}$, is given by

$$w_{p,th} = \frac{4\pi r^3 \rho_P \cdot N_{ion} \cdot \frac{c_{p,b}}{c_{i,b}} \cdot \left(\frac{i_{tr}}{i}\right)^\alpha \cdot H \cdot P_{(k/K,i)}}{\frac{3 M_m i}{zF} + 4\pi r^3 \rho_P N_{ion} \cdot \frac{c_{p,b}}{c_{i,b}} \cdot \left(\frac{i_{tr}}{i}\right)^\alpha \cdot H \cdot P_{(k/K,i)}} \cdot 100 \qquad (2)$$

where $c_{p,b}$ and $c_{i,b}$ are the bulk electrolyte concentration of particles and ions, respectively, N_{ion} is the total number of ions adsorbed on one particle. $P_{(k/K,i)}$ is the probability of particle incorporation accounting for the precondition of a specific residence time. In order to take into account hydrodynamic effects, an empirical factor H was introduced which is equal to unity under laminar flow and decreases to zero for the case of highly turbulent flow. Since the model assumes that particles and ions move at the same rate, the factor $(i_{tr} / i)^\alpha$ was introduced. Thereby, i_{tr} refers to the transition current density from charge transfer to mass transport control and determines the position of the experimentally observed maximum in the amount of incorporation vs. current density. The superscript α is equal to zero if $i<i_{tr}$ for kinetic limited condition and nonzero if $i>i_{tr}$ for mass transfer control. Similar to Guglielmi's model, this model cannot predict the layer composition from the process conditions. Moreover, the factors K, H, k, i_{tr} and α need to be determined by fitting the model with experimental results.

1.3.3 Trajectory Model of Fransaer et al.

The two models described so far have been developed for the ECD of Brownian particles, *i.e.* particles with a diameter below 1 µm. Since most industrial processes have utilized much bigger particles, Fransaer et al. developed a particle trajectory model to describe the codeposition of non-Brownian particles on a rotating disk electrode [55]. The model was based on a complete analysis of all forces and torques acting on the particle in suspension. This includes the forces due to fluid convection and particle motion as well as the external forces acting on the particle such as gravitational, electrophoretic, and double layer forces. The particle transport to the electrode was found to be mainly governed by convection and by the dispersion force, while the impact of the double layer force as well as electrophoresis can be neglected. The particle volume flux to the electrode can be determined by calculating the limiting particle trajectory which implies a separation of the particle trajectory of particles reaching the electrode from those passing by. However, this model fails close to the electrode surface because it leads to the "perfect sink" condition, which implies that all particles approaching within a certain critical distance of the electrode surface become irreversibly and instantaneously incorporated in the metal matrix. Hence, the authors implemented the particle-electrode interaction term, introducing a new equation for the probability of particle incorporation. The probability that a particle that is brought in contact with the electrode co-deposits, depends on the force balance on the particle in the vicinity of the electrode (Fig. 5). A precondition for particle incorporation in the growing metal film is fulfilled when the adhesion force between particle and electrode, F_{adh}, is greater than the shear force, F_{shear}. Hence, the trajectory model resulted in a good description of the variation of particle incorporation as a function of the bath particle concentration and the electrode rotation rate.

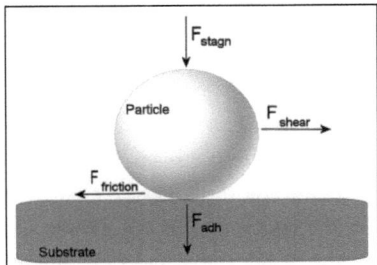

Figure 5 Forces acting on a rigid spherical particle in the vicinity of the electrode [55].

In order to describe the maximum in the amount of particle incorporation versus current density, the structural or hydration force between particle and electrode was introduced. This force, which is strongly repulsive at short distances, is due to the ordering of solvent molecules in concentrated electrolytes. The impact of the hydration force was expected to be highly dependent on the electrode potential and therefore on the applied current density. Hence, it was assumed that the maximum in the amount of particle incorporation versus current density curve could be caused by changes in the ordering of water dipoles at the electrode surface.

From all models discussed so far, the trajectory model possesses the uniqueness that it does not contain any adjustable parameter. However, the model has also its shortcomings, since it does not predict a maximum particle concentration with respect to current density. Additionally, it is not applicable to nanosized particles where Brownian motion is expected to dominate the particle motion.

1.3.4 Kinetic Model of Vereecken et al.

A model was presented by Vereecken et al. [51, 114] that took into account the kinetics and residence time of nanoparticles at the electrode surface. Therefore, they analyzed the nanoparticle flux during ECD via the forces acting on a single particle. Using a rotating disk electrode a kinetic model was achieved to describe the codeposition of 50 nm as well as 300 nm Al_2O_3 nanoparticles with nickel from a sulfamate electrolyte. This model took into account the convective diffusion of particles to the electrode surface and the effect of particle gravitational force on the particle flux. It was found in experiments that gravity would have as much as 40% effect on the incorporation of 300 nm particles. However, gravity had a negligible influence on the codeposition of 50 nm particles.

From Fick's first law and the Nernst diffusion layer model, the number of co-deposited particles per unit volume in the metallic matrix, d_p, was found to be

$$d_p = \frac{zF}{V_{m,M} i} \cdot \left(0.62 D_p^{2/3} \upsilon^{-1/6} \omega^{1/2} - \frac{\rho_p g r^2}{9\eta} \right) \cdot n_p \qquad (3)$$

where n_p is the particle density in the electrolyte, υ is the kinematic viscosity of the electrolyte, $V_{m,M}$ is the molar volume of the metal film, r is the particle radius, ω is the rotational speed of the electrode, η is the viscosity of the solution, D_p is the diffusion coefficient of the particle in the solution, ρ_p is the density of the particle, and g is the acceleration due to gravity. The model only applies for diffusion limited depositions on a rotating disk electrode (RDE).

1.3.5 Model of Lee and Talbot

Recently, Lee and Talbot [115] developed a model to describe the experimental results from the ECD of alumina nanoparticles with copper using a rotating cylinder electrode (RCE) adapting the models of Celis et al. (section 1.3.2) [1] and Vereecken et al. (section 1.3.4) [51, 114]. In this model the basic approach of Vereecken et al. [51, 114] was modified by replacing the transport parameters of the RDE by those of the RCE system. Moreover, the dependency of the flux of particles to the electrode on the current density was described by the concepts presented by Celis et al. [1], *i.e.* a distinction between charge transfer and mass transfer control was implemented. Finally, this model can be used to predict the amount of particle incorporation in the kinetically as well as in the mass transfer limited region of the ECD. Nevertheless, a few experimental data points are essentially to quantify the value of certain fitting parameters.

1.3.6 Summary

The work done over the last two decades has led to an improved understanding of the macroscopic effects of important process variables such as particle size, particle type, current density, hydrodynamics, electrolyte composition, etc. However, there is still a lack of fundamental understanding of the entrapment mechanism. The models proposed so far are still unable to predict the process conditions necessary to achieve a given film composition for a given system without an initial set of experiments to find the empirical fitting parameters used in the model.

1.4 Process Variables

Based on recent research on ECD, the important process parameters that influence the composition of the deposit have been identified as, but are certainly not limited to, the electrode orientation, hydrodynamics, particle characteristics such as size, shape and type, current density and current modulation, composition, temperature and pH of the electrolyte, and the particle loading of the electrolyte [2, 3, 33, 50, 52-54, 77, 78, 116, 117]. The influence of a particular variable on the ECD process is typically assessed by the change in the amount of particle incorporation obtained when that parameter is varied. It has to be emphasized that the effects of certain process parameters, which are often interrelated, may vary for different particle-metal combinations [2, 3, 116, 117]. The following section describes the effects of selected process parameters on the ECD process.

1.4.1 Electrochemical Cell Configuration and Electrode Orientation

In general, there are three main cell geometries that have been used to produce MMNC, *i.e.* the parallel plate electrode (PPE), the rotating disk electrode (RDE) and the rotating cylinder electrode (RCE). Most studies on ECD employ a PPE configuration for reasons of simplicity [2, 50]. However, due to the various ways of agitating the suspension, an analysis and comparison of hydrodynamics is rather difficult with this configuration. Another disadvantage of the PPE design is that the hydrodynamic and concentration boundary layer thickness are not constant over the electrode surfaces, especially at the electrode edges. This results in a variation of the concentration of reactants yielding a possibly non-uniform film composition. A better control of the hydrodynamic conditions can be achieved using a RCE [58-60] or RDE [21, 32, 60]. A detailed description of the flow field around these electrodes can be found in Refs. [102, 118, 119]. Disadvantages of the RDE configuration include a variation of mass transfer and current distribution as a function of the radial position, except at limiting current density, and of shear [102]. The benefit of the RDE is that it operates predominantly under laminar flow conditions. In contrast, the RCE system is characterized by a constant shear rate and a uniform current density distribution along the cylindrical electrode surface. However, the main disadvantage of an RCE is that the flow becomes turbulent even for low rotational rates [118]. In addition, both the RCE and RDE configurations are not typically a viable industrial method because of their limitation of specific electrode shapes.

Nowadays, the impinging jet electrode (IJE) system becomes more and more important due to its favorable advantages, viz. selective [62] and high-speed plating [63]. IJEs are particularly promising since plating rates can be up to 10 times faster compared to conventional systems [64]. Furthermore, jet plating is an attractive method to electrodeposit gradient coatings while controlling the volume fraction of particles by changing the jet velocity [64]. As long as the nozzle is placed close to the substrate, the electrodeposition is hydrodynamically limited to an area proportional to the width of the nozzle [120]. Hence, an IJE can be used to control hydrodynamics during deposition and increase the mass transfer [121]. Nevertheless, there are also several disadvantages of the IJE system particularly with regard to the ECD process. Due to the variety of permutations of the IJE cell geometry and orientation (*i.e.* nozzle size and shape, submerged/unsubmerged jet, angle of contact between fluid jet and electrode, etc.) obviously fewer hydrodynamic and mass transfer correlations are available for a particular IJE configuration compared to the RDE and RCE setup. First of all, it is important to distinguish between submerged and unsubmerged jet systems. In the case of a submerged IJE, the nozzle is placed inside the electrolyte with no air gap between the nozzle and the substrate. On the other hand, in an unsubmerged system the nozzle is placed so that there is an air gap between nozzle and substrate [122]. The main benefit of the unsubmerged system is to avoid the entrainment of the bulk fluid found in the submerged configuration which is particularly important for ECD to enable nearly homogenous particle entrainment in the fluid jet [122].

1.4.2 Hydrodynamics, Magneto-hydrodynamics

The previous section was partially related to the subject hydrodynamics. However, the main focus was directed to the experimental setup of the plating apparatus. The present section is targeted on a more detailed discussion how the hydrodynamic conditions can affect the particle transport to the electrode and the process of particle incorporation.

Electrolyte agitation is generally necessary in ECD to maintain the particles in suspension and to transport the particles to the electrode surface [2]. Hydrodynamics is an important factor since it controls the rate, direction and force with which the particles arrive at the electrode. Increased agitation has been found to increase particle incorporation by keeping the particles suspended in the electrolyte and improving the particle transport to the electrode [2, 95]. On the other hand, increased agitation also decreases the residence time of a particle at the electrode and increases shear which may sweep away loosely adsorbed particles [55, 60]. Hence, the particle-electrode contact is a necessary (section 1.3) but not at all a sufficient condition for particle incorporation [55]. Because of these competing tendencies, many ECD systems show maxima in particle incorporation [33]. Moreover, hydrodynamics can also affect the limiting current density of the metal deposition by altering the thickness of the diffusion layer [102].

The effect of the three hydrodynamic flow regimes, *i.e.* laminar, transitional and turbulent flow, on the particle incorporation was shown using a RDE system [123]. In the laminar flow regime, incorporation was found to remain constant regardless of the rotational speed. Transitional flow was characterized by a steady decrease in incorporation followed by a small peak at the transition to turbulent flow. In the turbulent flow regime, further increases in rotational speed only decreased incorporation [123]. The influence of shear and centrifugal force on ECD was evaluated by comparing experimental results of a RCE and RDE system under equivalent flow conditions [60]. It has been observed, that while shear may be the main cause of particle detachment, centrifugal force may influence incorporation by determining whether detached particles are transported away from the electrode. The impact of the centrifugal force increases with increasing difference in specific weight of the particles and electrolyte [60].

Furthermore, the presence of a static magnetic field during the codeposition process has been shown to have an impact on the composition and structure of the plated magnetic composites [124]. A static magnetic field can affect the ECD of magnetic nanoparticles with Ni in two different ways. On the one hand it influences the rate of mass transport via the magnetohydrodynamic (MHD) effect [125, 126]. The MHD effect, which is mainly caused by the Lorentz force f_L, induces fluid motion and thus can enhance the mass transport [125-127].

$$\mathbf{f}_L = \mathbf{j} \times \mathbf{B} \qquad (4)$$

Furthermore, the magnetic nanoparticles will move in the plating electrolyte towards regions with higher magnetic flux density, due to the magnetophoretic force (Eq. 5) [128].

$$\mathbf{f}_{mp} = \frac{\Delta \chi}{\mu_0} (\mathbf{B} \cdot \nabla) \mathbf{B} \qquad (5)$$

In the above equations j is the current density, B is the magnetic flux density (bold faces symbolize vector quantities), χ the magnetic susceptibility of the particles and μ_0 the magnetic permeability of free space. Notice that f_L acts on fluid elements whereas f_{mp} acts on the particles themselves. An increased incorporation of magnetic particles is expected if the working electrode is located normal to the trajectories of the particles.

1.4.3 Bath Properties

Many properties of the bath have been found to influence ECD, including particle characteristics and loading in solution, electrolyte composition and concentration, pH, particle-electrolyte interactions, additives, and temperature [2, 3, 21, 33, 129]. These effects will be discussed in more detail in the two following sections.

1.4.3.1 Electrolyte Composition

The electrolyte is composed of the metal cation undergoing cathodic deposition, its counter ion, additives, and either hydroxide ions or hydronium ions, depending on the working pH of the bath [130, 131]. The two key functions of the electrolyte are to furnish the ions that are reduced at the cathode and to minimize ohmic resistance across the cell [130, 131]. A variety of electrolytes have been used for the ECD process to deposit a metal matrix of Au, Ag, Co, Cu, Fe, Ni, or Cr, or their respective alloys [2]. Electrolytic deposition of a nickel matrix is typically performed using a Watts' bath which consists of nickel sulfate, nickel chloride and boric acid, and plating baths based on nickel acetate, fluoroborate or sulfamate [83, 132]. Nickel films deposited from the sulfamate bath feature low contents of sulphur and carbon incorporation combined with improved mechanical properties such as low tensile stress and increased hardness [132, 133].

As mentioned in section 1.3.2, some models for ECD assume that the particles are incorporated when a certain fraction of the metal ions adsorbed on the particle surface are reduced. Since adsorption processes are usually described by equations similar to the Langmuir isotherm, the adsorption of metal ions on the particle surface also depends on their bulk concentration in the electrolyte [134]. However, once the particle surface becomes saturated with ions, increasing the bulk ion concentration further should neither influence the concentration of adsorbed metal ions nor the ECD process. Furthermore, the adsorption of reactive metal ions onto the particle surface is not a necessary precondition for incorporation, as hydrophobic particles such as polystyrene [21] as well as negatively charged particles (due to the adsorption of other bath constituents such as citrate and pyrophosphate anions) [16, 17] are readily incorporated in a Cu or Ni matrix.

The electrolyte pH is important in determining overall amounts of incorporation [16, 17, 57] by affecting the particle surface charge (section 1.2). An increase in incorporation of diamond and alumina nanoparticles was found with increasing pH [16, 17, 57]. While there is no final explanation for this behavior, a tentative electrostatic model was proposed to explain qualitatively the relation between the surface charge of the particles and the amount of incorporated particles [16, 17, 57].

1.4.3.2 Particle Characteristics

Particles influence the ECD process via chemical composition, crystallographic phase, size, density, hydrophobicity, and concentration in the plating electrolyte [2, 3, 21, 33, 129]. Particle composition is extremely important in determining the overall amount of particle incorporation [2]. For example, approximately three times more TiO_2 than Al_2O_3 have been incorporated into nickel under the same deposition conditions [2]. In particular, hydrophobic particles such as polystyrene (PS) and SiC typically incorporate more readily than hydrophilic particles such as metal oxides [129, 135]. The forces between either PS or SiO_2 particles and the surface of a copper electrode before, during, and after electrocodeposition have been examined using an atomic force microscope [135]. The improved incorporation of hydrophobic particles was found to be due to attractive hydrophobic and van der Waals interactions between the PS particles and the copper electrode [135]. In contrast, the hydrophilic SiO_2 particles were found to be separated from the electrode by a thin film of electrolyte due to the repulsive hydration force [135]. ECD is influenced not only by the composition of the particles, but also by their crystallographic phase. While one phase may readily incorporate into the metal matrix, another phase with the same chemical composition may only incorporate to a small extent or not at all [2]. Typically, ECD was studied using particle concentration in the electrolyte between 1 and 200 g l^{-1} [2]. In most cases, increasing particle loading resulted in an increase in particle incorporation [2]. However, particularly at higher loadings, the increase in incorporation is not proportional to the increase in loading [2].

1.4.4 Current Density and Current Modulation

Current density has an important, but poorly understood, effect on the ECD process [2, 3]. Particle incorporation was found to decrease with increasing current density and then levels off for the systems Ni-diamond and Ni-Cr [136-139]; but also an additional effect at lower current densities was observed, where an increase of current density increases particle incorporation for the systems $Ni-TiO_2$, $Cu-Al_2O_3$, $Cr-Al_2O_3$ and Co-SiC [1, 2, 58, 81, 95, 140]. An important observation is that the maximum in particle incorporation depends considerably on both, the current density and the hydrodynamics of the system due to the effect of hydrodynamics on the current-potential relationship [58, 95, 131]. Interactions between hydrodynamics and current density on the ECD process can be accounted by normalizing the current density with the limiting current density [58, 95, 138].

A variety of galvanostatic techniques have been utilized for ECD investigations, including direct current (DC), pulsed plating (PP) and pulsed reverse plating (PRP) [3]. Both, PP [54, 74-76] and PRP [26, 77] were found to improve particle incorporation and MMNC film properties. Using PRP, the maximum incorporation of alumina particles in a Cu matrix was found when the deposit thickness plated per cycle approaches the particle diameter [77]. In the case of $Ni-Al_2O_3$ composites, PP had led to an incorporation of smaller and less agglomerated nanoparticles [79, 80].

1.5 Structure and Properties of Metal Matrix Nanocomposites

The effects of particle incorporation on the microstructure and physical properties of MMNC films have been extensively discussed in the literature. Several studies have found an obvious increase in hardness compared to the pure nickel film due to the reinforcement with various particles such as TiO_2, Al_2O_3, and SiC. The properties of MMNCs mainly depend on their composition and structure [11]. Uniform distribution and high amounts of incorporated particles within the metal matrix were found to be crucial to improve the film properties [16, 67]. In many cases the enhanced performance of the composite film is mainly caused by a change in the growth mode or grain size of the metal matrix and not so much by the presence of the particles themselves [16, 30].

In general, there are two common mechanism of grain refinement of metal films. The first method is the particle incorporation in a electrodeposit [3, 15, 76]. In this regard, grain boundary pinning is generally agreed upon in the literature as the mechanism of grain refinement [76, 129]. Even though the particles are not coherent with the metal matrix, they inhibit the movement of dislocations due to the physical interaction between the particle reinforcement and the dislocations. With small impenetrable particles, a passing dislocation will bow between particles and finally pass by leaving behind an so-called Orowan loop [141]. By increasing the density and length of dislocations, the presence of these small particles tends to improve the mechanical properties of the material sustainable [141]. On the other hand, pulse plating can be used for further grain refinement [69]. Enhanced mass transfer during PP can be attributed to the grain refinement due to creation of nucleation sites at the beginning of each cycle [69].

While mechanical properties and corrosion resistance of composite properties can be greatly enhanced [142] some material properties, *e.g.* magnetism and electrical resistance, are sacrificed in the MMNC film [16, 80].

2 Experimental

The electrolyte composition (section 2.1.1), experimental setup (section 2.1.3), substrate preparation (section 2.1.4), and techniques used for characterization of MMNC films (section 2.3) are described in this chapter. Prior to the ECD experiments, an extensive characterization of the utilized nanoparticles (section 2.1.2), regarding to size, shape, surface charge and sedimentation behavior in the plating electrolyte, is required in order to gain insight in the composite film formation. The particular techniques used are described in the section 2.2.

2.1 Deposition Procedures

2.1.1 Electrolyte Composition and Working Conditions

The ECD experiments were carried out using either an acidic sulfamate or an alkaline pyrophosphate electrolyte. The composition of the two nickel baths and the operating conditions employed for plating are shown in Table 1. All chemicals were p.a. grade and doubly distilled water was used for the preparation of all solutions. Suspensions were prepared by adding a specified amount of nanoparticles (section 2.1.2) to the electrolyte. The dispersion was stirred using a magnetic stirrer for 12-24 h in a covered beaker. The pH of the solution was measured before and after each experiment and if necessary adjusted to pH 4.3 or 9.5 with either $NaHCO_3$ or H_2SO_4. Apart from the IJE experiments and those in the magnetic field which were performed at room temperature, the electrolyte temperature was maintained at 40 °C using a Haake thermostate (model G D1, accuracy ± 1°C).

Table 1 Electrolyte composition and working conditions.

	Sulfamate bath	Pyrophosphate bath
Nickel	1.08 M $Ni(NH_2SO_3)_2$	0.30 M $NiSO_4 \cdot 6\ H_2O$
	0.04 M $NiCl_2 \cdot 6\ H_2O$	
Other	0.65 M H_3BO_3	0.75 M $K_4P_2O_7$
		0.12 M $C_6H_8O_7 \cdot H_2O$
pH	4.3	9.5
Temperature [°C]	25-40	25-40
Current density [A dm^{-2}]	1-30	

2.1.2 Nanoparticles

2.1.2.1 Commercial Ceramic Nanoparticles

The alumina nanoparticles used for the ECD experiments were commercially available with a primary particle size of either 13 nm (Aeroxide Alu C, Degussa) or 50 nm (Buehler Micropolish II) as given by the producer. The TiO_2 particles (Aeroxide TiO_2 P 25, Degussa) are specified by a mean diameter of 21 nm. Further details on the particle properties are given in Table 2. All particles were used as received without any further treatment.

Both, the 13 nm Al_2O_3 and the 21 nm TiO_2 particles are produced by the AEROSIL® process which implies the high-temperature hydrolysis of the particular gaseous metal chlorides [143]. The resulting powder is a very fine fumed metal oxide with a high specific surface area. Crystallographically, alumina Alu C has to be assigned to the δ-group, a subgroup of the thermodynamically unstable γ-modification which is formed at temperatures between 750 and 1000° C [144]. The γ-modification is characterized by a cubic closest sphere packing of the O^{2-} ions, at which the octahedral and tetrahedral lattice vacancies are statistically occupied by the Al^{3+} ions [144]. In the titania P25, the anatase modification dominates by about 70% over the thermodynamically more stable rutile form. Anatase and rutile have the same tetragonal symmetry, despite having different structures [144]. In the rutile form, the Ti^{4+} ions occupy half of the octahedral holes in a hexagonal closest packing of the O^{2-} ions. In contrast to rutile, anatase is characterized by a cubic closest packing of the O^{2-} ions, at which half of the octahedral holes are statistically occupied by the Ti^{4+} ions [144].

The 50 nm γ-Al_2O_3 particles were distributed by Buehler, but manufactured by Praxair via precipitation from an aluminum ammonium alum, followed by a heat treatment step.

Table 2 Particle properties as given by the supplier [143, 145].

Particle Property	Unit	Aeroxide Alu C Al_2O_3	Micropolish II Al_2O_3	Aeroxide P25 TiO_2
Average primary Particle Size	nm	13	50	21
Specific Surface (BET)	$m^2\ g^{-1}$	100 ± 15	/	50 ± 15
Density	$g\ cm^{-3}$	3.2	3.84	3.7
Purity	%	>99.6	/	>99.5

2.1.2.2 Synthesis of Magnetic Nanoparticles

Preparation of cobalt nanoparticles

The acicular Co nanoparticles in the shape of cubes and discs were synthesized by Dr. Christian P. Gräf (Saarland University, Department of Engineering Physics) by thermal decomposition of the precursor di-cobalt octa-carbonyl, $Co_2(CO)_8$. Due to their anisometric shape they have a high magnetic hardness. A detailed description of the synthesis route as well as an extensive characterization of those particles is given in Ref. [146]. For further use the Co nanoparticles were washed several times with anhydrous ethanol and hexane.

Preparation of magnetite nanoparticles

Magnetite (Fe_3O_4) particles were prepared by chemical precipitation from a ferric and ferrous salt solution upon addition of NaOH according to the procedures described in Refs. [147-151]. All chemicals (p.a. grade) were purchased from Fluka. In a typical procedure, 2.35 g anhydrous $FeCl_3$ and 1.43 g $FeCl_2 \cdot 4H_2O$ were mixed in 40 ml doubly distilled deoxygenated water, yielding the required molar ratio of 2:1 for Fe^{3+} and Fe^{2+}. The resulting pH-value was about 1.2. This mixture was filled in a two necked flask, equipped with a reflux-cooler and a septum. The deoxygenation was achieved by bubbling nitrogen through the respective solutions for at least 30 minutes. Under continued nitrogen bubbling, the mixture was heated to 80°C. While vigorously stirring the reaction mixture, 50 ml of 1.5 M deoxygenated NaOH was quickly injected through the septum, which caused an increase of the pH-value to 11. The solution immediately turned black. After 30 minutes continued heating, 1 g of citric acid dissolved in 2 ml of 1.5 M NaOH was introduced. The pH decreased to 10.3 and another 4 ml of 1.5 M NaOH were added to reach pH 11 again. After that, the temperature was increased to 95°C, and stirring continued for another 90 minutes. After cooling the reaction solution to room temperature, the particles were magnetically sedimented using a permanent magnet, washed with water and re-suspended by sonication. This cleaning step was repeated two times.

Direct images of the synthesized cobalt and magnetite nanoparticles were obtained by transmission electron microscopy (TEM) using a JEOL JEM-2010 microscope at an acceleration voltage of 200 kV in the bright field image mode. Samples for TEM imaging were prepared by evaporating a drop of the highly diluted dispersion on a carbon-coated copper grid.

2.1.3 Experimental Set-up

2.1.3.1 Parallel Plate Electrode (PPE)

The experimental setup of the parallel plate electrode system is shown in Fig. 6. A double-walled glass cell with a volume of 300 ml was used to perform the experiments. The temperature was controlled with a Haake thermostat (model G D1, accuracy ± 1°C). The plating solution (composition and working conditions as listed in Table 1, p. 22) was mechanically stirred (250 rpm) using a magnetic stirrer. In the three electrode system, the anode was a cylindrically bent nickel plate with an area of about 120 cm². A saturated calomel electrode (SCE KE10, KSI Meinsberg, Germany) was used as the reference electrode.

The direct current (DC) deposition was carried out with a potentiostat/galvanostat model PS6 (Sensortechnik Meinsberg GmbH, Germany) or model 273 (EG&G Princeton Applied Research) with a current density between 1 and 10 A dm^{-2}. The total plating charge was about $45 \cdot 10^2$ C dm^{-2} for each deposit yielding deposits ca. 17 µm thick.

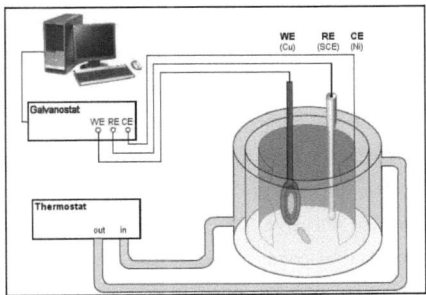

Figure 6 Experimental setup of the parallel plate electrode system.

Pulse plating (PP) and pulse reverse plating (PRP) experiments were carried out using a CAPP pulse-plating system (Axel Akerman S.A., Denmark). In the PP experiments the pulse frequency [i.e. $f_p = 1 / (t_{on} + t_{off})$] and the duty cycle [i.e. $\theta = t_{on} / (t_{on} + t_{off})$] of the imposed rectangular pulses was varied. The pulse on-time (t_{on}) was varied between 25 and 200 ms and the pulse pause (t_{off}) was kept constant at 100 and 200 ms, respectively as shown in Table 3. For the PP experiments the cathodic peak current density (i_p) was maintained at 5 and 10 A dm^{-2}, respectively. The overall ECD times were chosen so that the total charge was always $62 \cdot 10^2$ C dm^{-2}. This corresponds to a layer thickness of about 21 µm for the pure nickel assuming 100 % current efficiency. In the case of PRP depositions the pulse on-time (t_{on}) was varied between 50 and 400 ms and the pulse reverse time (t_{rev}) was kept constant at 20 ms. The cathodic peak current density (i_p) was 5 A dm^{-2} whereas the anodic peak current density (i_{an}) varied between 1 and 5 A dm^{-2}.

Table 3 Applied pulse plating current programs. The cathodic peak current density was 5 and 10 A dm^{-2}, respectively.

t$_{off}$ / ms	t$_{on}$ / ms	f$_P$ / Hz	θ / %
25	100	8.0	20.0
50	100	6.7	33.3
100	100	5.0	50.0
25	200	4.4	11.1
50	200	4.0	20.0
50	100	3.3	66.7
200	200	2.5	50.0
400	200	1.7	66.7

2.1.3.2 Impinging Jet Electrode (IJE)

The unsubmerged IJE system used was designed, built, and tested by Steve Osborne (Chemical Engineering, University of California San Diego) [122]. A brief description of the optimized apparatus will be provided below. Further details related to the development and design of the apparatus may be found in Osborne's PhD thesis [122]. The selection of the IJE system components was based on the major requirements of corrosion and wear resistance due to the acidic metal plating baths and a high concentration of sub-micron abrasive particles. Also, the ability to re-circulate the electrolyte solution is essential to conserve materials and minimize waste. In addition, a constant particle loading must be maintained throughout the apparatus and during the ECD experiments. Hence, it is essential to ensure adequate mixing and to avoid particle buildup as the electrolyte is re-circulated.

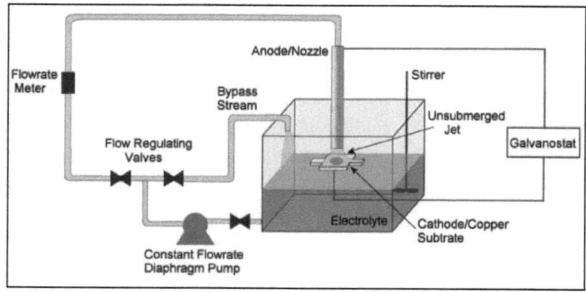

Figure 7 Experimental setup of the impinging jet electrode system.

Figure 7 shows a schematic diagram of the jet electroplating system, which consist of two main components, *i.e.* systems for controlling hydrodynamics and for controlling the charge passed through the apparatus. The electroplating cell was directly placed over the 8 l rectangular solution tank (Saint-Gobain 8 L HDPE tank). An electric diaphragm pump (Jabsco model 31801-0115) with an 8 l min^{-1} capacity was used to circulate the electrolyte. The flow rate was accurately adjusted with two ball valves that divided the flow into a main and a bypass stream. Flow rate was measured with a digital paddle flow meter (Blue-White Industries, F-1000-RB). The particles were kept in suspension using a laboratory stirrer (Fisher Scientific, model no. 47) with a paddle impeller. The electrolyte was continuously recirculated to the stirring bath. A Plexiglas cover prevented splashing and excessive evaporation, in addition to supporting the impinging jet cell assembly.

The electroplating cell of the IJE system is schematically shown in Fig. 8. A 1.1 cm inner diameter stainless steel tube served as anode and nozzle. The nozzle was held into place above the substrate using two plastic rods. The nozzle-substrate gap distance was 0.65 cm and the angle at which the jet impinges onto the surface was 90°.

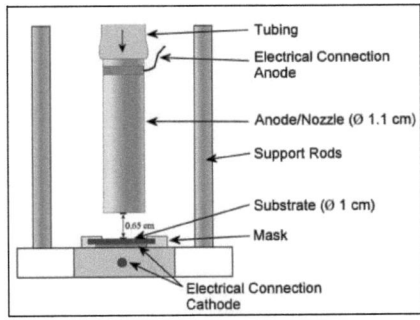

Figure 8 Schematic sketch of the electroplating cell of the IJE system.

ECD experiments were conducted at room temperature with an acidic sufamate electrolyte and the working conditions as listed in Table 1 (p. 22). A copper plate (2 x 2 x 0.15 cm) was employed as substrate. The plating area was a 1 cm diameter circle cut into electroplaters' masking tape (3M) which covered the rest of the substrate area. The codeposition was carried out galvanostatically with a galvanostat (model 273, EG&G Princeton Applied Research). The process variables investigated were current density (5-30 A dm^{-2}), electrolyte flow rate (1-6.5 l min^{-1}), and particle loading (10 to 120 g l^{-1}). The total charge was about 12.9·10^3 C dm^{-2} yielding deposits ca. 40 µm thick. After electrolysis, the samples were sonicated in water for ten minutes to remove loosely adherent particles from the surface.

2.1.3.3 Electrocodeposition in a Magnetic Field

A three electrode setup in a lab made PTFE cell (volume ca. 10 ml) was used for the depositions in a static magnetic field. The magnetic nanoparticles (section 2.1.2.2) were suspended in an alkaline nickel pyrophosphate bath (composition and working conditions as listed in Table 1, p. 22). The particle content of the electrolyte varied between 0 and 10 g l^{-1}. Copper discs with an electroactive area of ~0.22 cm^2 were used as substrates. A saturated calomel electrode (KSI Meinsberg, Germany) was used as reference electrode and a platinum plate as anode. The ECD experiments were carried out galvanostatically with a potentiostat/galvanostat model 263A (EG&G Princeton Applied Research) at room temperature with current densities between 0.5 and 10 A dm^{-2}. The total plating charge was about 18·10^2 C dm^{-2} for each deposit. Assuming 100 % current efficiency this would correspond to a layer thickness of about 6 μm. During codeposition the plating electrolyte was agitated by bubbling N$_2$ in order to prevent sedimentation of the particles. After plating, the samples were rinsed with deionized water, sonicated for ten minutes to remove loosely adherent particles and dried in air at room temperature.

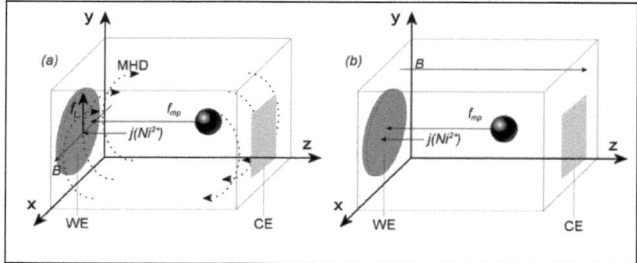

Figure 9   Schematic sketch of the experimental setup of the electrocodeposition in a (a) parallel (x-direction) and (b) perpendicular (z-direction) magnetic field (**B** field). WE working electrode, CE counter electrode. The black sphere represents the particle and **j**(Ni^{2+}) is the electric current carried by the Ni deposition. **f**$_L$ and **f**$_{mp}$ are the Lorentz and magnetophoretic force, respectively.

A water-cooled electromagnet (VEB Polytechnik, Phylatex, Germany) was used to apply a magnetic field of 100 mT. The magnetic flux density was adjusted by varying the electric current through the coils and measured by a Hall probe (Lake Shore, model 450). Further details on the setup can be found in Ref. [124, 152]. The magnetic field in the air gap of the electromagnet had an inhomogeneity of some mT cm^{-1}. However, with the cell inserted into the gap and due to the growth of a ferromagnetic layer at the working electrode the field becomes inhomogeneous. In the following discussion, the terms perpendicular (z- direction) and parallel (x-direction) refer to the orientation of the **B** field relative to the working electrode (see Fig. 9 for details).

2.1.3.4 Electrochemical Quartz Crystal Microbalance (EQCM)

The adsorption behavior of alumina nanoparticles onto the electrode was investigated using the electrochemical quartz crystal microbalance (EQCM) [131, 153]. In order to obtain controllable hydrodynamics a submerged impinging jet cell was used [154-156].

In the case of rigid layers with a uniform thickness, the mass change per area at the quartz crystal surface, $\Delta m/A$, can be calculated from the shift of the resonance frequency, Δf, using the Sauerbrey equation [157], Eq. 6.

$$\Delta f = -\frac{2 f_{R,0}^2 \Delta m}{\sqrt{\mu_Q \rho_Q}\, A} = -C_{SB} \frac{\Delta m}{A} \qquad (6)$$

where $f_{R,0}$ is the resonance frequency of the unloaded quartz (here $f_{R,0} = 10$ MHz), μ_Q and ρ_Q are the shear modulus and the density of quartz, respectively [154, 155]. The sensitivity of the EQCM is high, since a shift of 1 Hz in the resonance frequency of a 10 MHz quartz crystal corresponds to a mass change of 4.425 ng cm^{-2}.

For the in situ characterization of the particle adsorption process a dissipative EQCM technique (with damping monitoring) was used, which is based on an admittance spectrum of the quartz crystal near its resonance frequency. The resonance curve corresponds to a Lorentzian centered at the resonance frequency and with a full width at half maximum (FWHM) which is proportional to the damping of the quartz crystal. From the frequency shift one can get information about the mass deposited on the quartz, while the FWHM shift provides information on the damping of the quartz [154, 155] as well as on the surface roughness of the deposited layer. Since the adsorption of particles might yield non-rigid and porous layers, a precise monitoring of the damping of the quartz is necessary. An interpretation of the EQCM data by means of Eq. 6 is only feasible if the frequency shift is higher than the FHWM shift [154, 155].

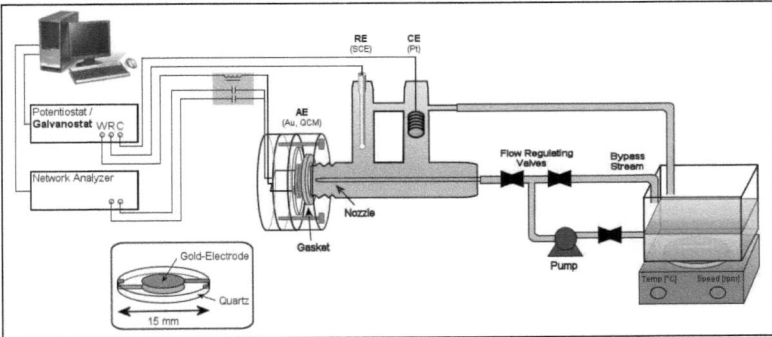

Figure 10 Schematic sketch of the submerged impinging jet cell used for the EQCM experiments.

Figure 10 shows the experimental setup of the submerged IJE system used in the particle adsorption measurements. There is a rectangular solution tank with a volumetric capacity of about 1 l. A magnetic stirrer is used to keep the particles suspended in the electrolyte. In the bottom of the electrolyte tank, there is an outlet for the solution to flow into the rotary pump. After the pump, a tee and two valves control the relative amounts of electrolyte flow through the nozzle and the bypass loop. The main loop subsequently impinges onto the quartz that is held into place with a special holder. The flow rate was adjusted manually in the range between 0.11 and 0.6 l min^{-1} by measuring the time taken to fill a vessel with a volume of 200 ml.

For the EQCM experiments, 10 MHz optically polished AT-cut quartzes (XA2579, Vectron Int. KVG, Germany) with gold electrodes (diam. 5 mm, thickness ca. 100 nm) on a ca. 5 nm chromium adhesion layer were used. These quartzes are approximately 15 mm in diameter and 165 µm thick as shown in the inset of Fig. 10. One gold electrode of the quartz crystal was used as the working electrode (active area ~0.22 cm^2), the reference was an Ag/AgCl electrode (SE11, KSI Meinsberg, Germany) and the counter electrode was a platinum coil. The quartz crystal was placed upright in front of the nozzle, and fixed with a silicone gasket (Fig. 10). An network analyzer (R3753BH, Advantest, Japan) with a pi network adapter (PIC-001, Advantest, Japan) was used to record the resonance spectrum of the quartz crystal from which the frequency and the damping shift were obtained [154, 155]. A potentiostat/galvanostat model 263A (EG&G Princeton Applied Research) was used for the electrochemical experiments. Decoupling of the high and the low frequency signals of the quartz and the electrochemistry was achieved by using a LC network (C = 0.1 µF, L = 1 mH). The measurement system was completely computer controlled using lab made control and evaluation software [155]. The process variables investigated were cathode potential of 0 to -1 V vs. Ag/AgCl, electrolyte flow rate of 0.11-0.6 l min^{-1}, and particle loading from 0 to 10 g l^{-1} Al$_2$O$_3$.

In order to make the particle adsorption testing comparable to the ECD experiments, particularly with regard to the ionic strength of the electrolyte solution (Table 1, p. 22), 1 M KCl was used as supporting electrolyte. Particle adsorption was studied on the gold electrode of the quartz as well as on a freshly prepared nickel film.

2.1.4 Substrate Preparation

The copper substrates used for the PPE system were mechanically polished with abrasive silicon carbide paper, electrochemically degreased in alkaline solution (UNAR EL 63, Schering), activated with uniclean 675 solution (Atotech Germany GmbH) and finally rinsed with doubly-distilled water before deposition. The substrates used for the IJE system were placed in an isopropanol bath and sonicated for 20 minutes.

2.2 Particle-Characterization

2.2.1 Zeta Potential Measurement

The zeta potential (ζ) of the nanoparticles was determined by measuring the mobility of charged particles in suspension under the influence of an electric field gradient [101, 105]. The surface of a colloidal particle dispersed in an electrolyte solution is characterized by an EDL as discussed in section 1.2.1 and shown in Fig. 2 (p. 7). Typically, ζ potential is defined as the potential difference between the shear plane and the bulk solution [101].

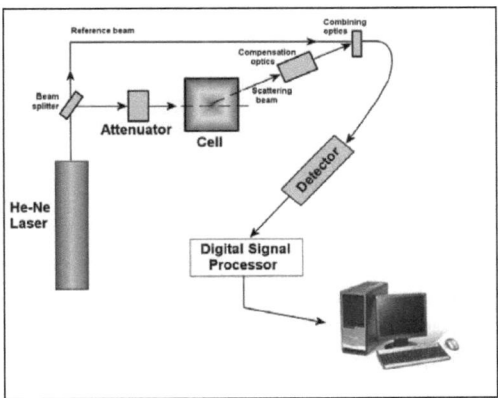

Figure 11 Schematic representation of the principle of the microelectrophoretic measurement [105].

The zeta potential measurement system comprises six main components: laser, cell, attenuator, detector, digital signal processor and a computer workstation as shown in Fig. 11. The laser (which is a 633 nm 'red' laser in our case) is used to illuminate the sample particles immersed in a fluid within the cell. The laser passes through the centre of the sample cell and the scattering is detected at an angle of 17°. The attenuator is implemented in order to optimize the intensity of scattered light arriving at the detector by reducing the intensity of the laser. This is essential, since the detector will become overloaded if too much light is detected. Usually, an appropriate attenuator position is determined automatically by the device. However, a manual fixing can be done by experienced users. During the measurement, any particles moving through the measurement volume will cause a fluctuation of the scattered light called Doppler frequency shift. By comparing the difference in frequency (*i.e.* the Doppler shift) between the scattered light and the incident light (*i.e.* reference beam), the mobility of the particles under the influence of the applied electric field can be determined [101]. In the case of electrical non-conductive, spherical particles with a thin double layer compared to the particle diameter, the Smoluchowski formula can be used to calculate the zeta potential from the measured particle mobility [101].

The zeta potential of the several nanoparticles was determined with a Zetasizer Nano (Malvern Instruments, Herrenberg) in 10^{-3} M KCl as well as in the diluted plating electrolytes with a concentration of 0.2 g l^{-1} nanoparticles in the pH range 2-11. The pH was adjusted with HCl or NaOH. The individual electrolyte components (Table 1, p. 22) were added in concentrations of 10^{-3} M. Dilute solutions of the electrolytes were prepared by mixing 0.02-0.1 ml of the corresponding electrolyte with 1 l of 10^{-3} M KCl. This dilution procedure is necessary to adjust the ionic strength of the solution to the requirements of the zeta potential measurement (section 1.2.1). Electrophoretic mobilities were converted to ζ-potentials using Smoluchowski's equation [101].

2.2.2 Photon Correlation Spectroscopy

Photon Correlation Spectroscopy (PCS) is used to determine the size of particles suspended in a solution [101]. The measuring principle is based on the Brownian motion of the particles which is mainly due to the random collision with the molecules of the liquid surrounding the particle. According to the Stokes-Einstein equation the speed of the particle due to Brownian motion is directly related to the particle size [101], whereupon small particles move more quickly compared to larger ones. Using this technique, particle sizes in the range from 0.6 nm to 6 µm can be determined [105].

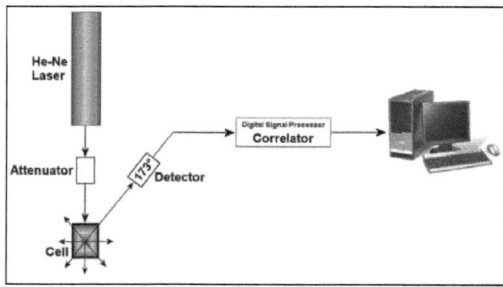

Figure 12   Schematic representation of the principle of the photon correlation spectroscopy [105].

Similar to the system described in the previous section, there are six main components of the PCS system used in this study: laser, cell, attenuator, detector, correlator and a computer workstation. A schematic representation of the PCS system is shown in Fig. 12 [105]. The function of the first three components, *i.e.* laser, cell and attenuator, is identical to that of the zeta potential measurement system. The colloidal particles scatter the light, and the scattering pattern depends mainly on the particle size and the wavelength of the laser beam [101]. In our case, the detection optics is arranged at a position of 173°. The benefits of the backscattered detection include the reduction of multiple scattering and the feasibility of measuring higher concentrations of the sample. Furthermore, the effect of contaminations such as dust is greatly reduced since large particles mainly scatter in the forward direction [105]. The scattering signal from the detector is then passed to the correlator. The correlator compares the scattering intensity at successive time intervals to derive the rate, at which the intensity is varying, what is called the autocorrelation function. Intensity variations are due to the fact that the scattering particle is moving when the photon hits it. Hence, the re-radiated light is characterized by a slightly different frequency, which is called Doppler broadening [101]. By fitting the autocorrelation function to an exponential function, the diffusion coefficient of the particles can be calculated. Based on the assumption of a spherical particle shape, the Stokes-Einstein equation can be used to determine the hydrodynamic particle diameter from the diffusion coefficient [101].

Finally, PCS is a fast technique, sensitive to nanosized particles, and providing information about the particle size as well as the size distribution. However, care must be taken when interpreting the size information since it is based on mathematical calculations which require a precise definition of the physical properties of the sample, viz. dispersant viscosity, temperature and refractive index [101]. Hence, electron microscopy should be considered to obtain visual and descriptive information about the nanoparticle population. However, electron microscopy does not yield quantitative information, *e.g.* particle size distributions. As a conclusion, a combination of a micrographic (electron microscopy) and a computational technique (light scattering) should be used together in the particle size characterization [101].

The size of the nanoparticles was determined with a Zetasizer Nano (Malvern Instruments, Herrenberg) in 10^{-3} M KCl as well as in the plating electrolytes with a concentration of 0.2 g l^{-1} nanoparticles in the pH range 2-11.

2.2.3 Dispersion Stability

The stability of the dispersed nanoparticles in the plating electrolytes was investigated by the sedimentation method using a special centrifuge (LUMiFuge 114, L.U.M. GmbH, Germany). The measuring principle [158] is based on an integrated optoelectronic sensor system which allows the recording of the spatial and temporal changes of light transmission during the rotation of up to 8 samples (Fig. 13). Throughout the measurement, transmission profiles are recorded and the sedimentation process is characterized by the time dependent displacement of the relative position of the boundary between supernatant and sediment. A resolution better than 100 µm can be achieved. Finally, the transmission profiles can be transformed into sedimentation-time curves.

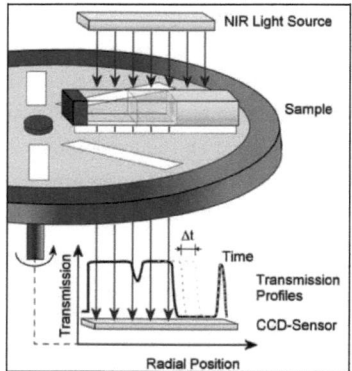

Figure 13 Schematic sketch of the measuring principle of the LUMIfuge 114.

Dispersion stability measurements of the several nanoparticles (section 2.1.2) suspended in the nickel electrolytes (Table 1, p. 22) were made in glass tubes at rotation rates between 300 and 1000 rpm. The slope of the sedimentation curves was used to calculate the sedimentation velocity and to assess the stability of the suspensions.

2.3 Layer-Characterization

Due to the fact that structure and properties are intimately related with the application of any material, the characterization of these properties is a very relevant issue in MMNC film development [159]. This chapter is concerned with the structural analysis of the deposited MMNC films by electron microscopy (sections 2.3.1.2, 2.3.2), X-ray diffraction (section 2.3.3) and spectroscopic procedures (section 2.3.1.3). While X-ray diffraction yields information from a large volume of the sample, electron microscopy probes a rather confined volume. Hence, both methods have to be considered of comparative importance for a complete characterization of structure and morphology of the deposited MMNC films. Additionally, the mechanical properties of the MMNC films were studied by microhardness (section 2.3.4) and abrasion testing (section 2.3.5), whereas the magnetic properties were examined using vibrating sample magnetometry (section 2.3.6). Further details about procedures for the utilized techniques will be provided in this section.

2.3.1 Particle Incorporation Analysis

The amount of co-deposited nanoparticles was determined by several techniques, including the electrogravimetric method (section 2.3.1.1), scanning electron microscopy (SEM) and energy-dispersive X-ray spectroscopy (EDX) (section 2.3.1.2), and glow discharge optical emission spectrometry (GD-OES) (section 2.3.1.3). The measuring principle of the several analyses is discussed in the following sections.

2.3.1.1 Electrogravimetric Analysis

The electrogravimetric analysis was employed to calculate the incorporation of alumina nanoparticles during nickel plating using the IJE system. The efficiency of pure nickel deposition from the sulfamate bath was determined within the investigated current density range. The theoretical mass of deposited nickel was calculated from Faraday's law and subsequently compared to the actual plated nickel mass. The current efficiency was in the range of 87-99 %. Hence, the amount of nickel electrodeposited during ECD was calculated from the total charge passed, taking into account the current efficiency of pure nickel plating. The typical weight of the co-deposited films was 0.1 g, resulting in an accuracy in the particle incorporation of ±0.6 vol-%. The amount of co-deposited alumina was determined by subtracting the amount of nickel determined from coulometry from the actual weight gain.

2.3.1.2 Scanning Electron Microscopy (SEM) and energy-dispersive X-ray Spectroscopy (EDX)

Scanning electron microscopy (SEM) can be used for qualitative surface morphology analysis; energy dispersive X-ray spectroscopic (EDX) is a technique to determine the chemical composition of a material non-destructive [141]. SEM is a type of microscope that uses an electron-beam rather than light to obtain an image of the sample. The advantages of SEM include a higher magnification, a larger depth of focus and a greater resolution in comparison to an optical microscope [141].

SEM micrographs combined with EDX studies were done in a DSM 982 Gemini (Zeiss Oberkochen, Germany) with an EDX detector model Voyager III 3200 (SUS Pioneer). The amount of co-deposited nanoparticles in the nickel matrix was evaluated from the aluminum (Al_2O_3), cobalt (Co), titanium (TiO_2) and iron (Fe_3O_4) signal, respectively. The particle concentration of the metal oxides was determined taking into consideration stoichiometric ratio of oxygen to metal in Al_2O_3, TiO_2 or Fe_3O_4. Three randomly chosen points each covering an area of 100 µm² were analyzed and an average value was calculated. The accuracy of EDX analysis was approximately ±0.5 vol-%. The EDX area analyses were performed both on the surface as well as in the cross section at a 10000x magnification. Surface morphology and microstructure of the deposited composites were also analyzed with SEM. For the cross section, the samples were embedded in epoxy resin and cut with a diamond saw. After mechanical grinding with 800 to 4000 grade silicon carbide paper and polishing with diamond suspension down to 1 µm, the cross-sections were etched approximately 10 seconds with a solution of 50 ml conc. CH_3COOH and 50 ml conc. HNO_3 [16].

A complete and quantitative representation of the sample microstructure can be established with electron backscattered diffraction (EBSD) [160, 161]. A FESEM NEON40EsB (Zeiss, Germany) was used at 25 kV with a scanning transmission electron microscopy (STEM) detector as well as an EBSD camera (EDAX TSL). For backscattered secondary electron (BSE) imaging, the voltage was lowered to 10 and 5 kV, respectively. Cross sections were prepared with a final oxide polish (OP). EBSD is a powerful microstructural-crystallographic technique used to investigate the crystallographic texture or preferred orientation of any crystalline or polycrystalline materials [161]. For this purpose an electron beam strikes a tilted sample and the diffracted electrons form a pattern which can be used to determine the crystal orientation, measure grain boundary misorientations, discriminate between different materials, and provide information about local crystalline perfection. The scanning and mapping capabilities of an EBSD system allow for rapid acquisition of data at sub-micron resolutions revealing the constituent grain morphology, orientations, and boundaries of the sample [160, 161].

2.3.1.3 Glow Discharge Optical Emission Spectrometry (GD-OES)

Glow Discharge Optical Emission Spectrometry (GD-OES) is typically defined as an instrumental technique which uses a low-pressure, obstructed, hollow-anode glow discharge combined with one or more optical spectrometers [162, 163]. In GD-OES analysis the sample is applied to a high negative potential of 700 V in a low pressure argon environment. The electrons and argon atoms interact to form a plasma. The collision of high-energy electrons with argon atoms causes the generation of Ar^+ ions. The positive argon ions then are driven by the negative bias to collide with the sample surface where they sputter (or erode) material uniformly [163]. Some of this sputtered material diffuses into the glow discharge plasma where it is dissociated into atomic particles and finally excited. The light emitted from these excited state species, as they collapse back to lower energy levels, is characteristic of the elements composing the sample. The wavelengths and intensity of the light emission are used to identify and quantify the composition of the sample. The aim of GD-OES analysis is most often the determination of the concentration of an element as a function of depth. Therefore, the initial measured information, *i.e.* the intensity vs. time relation, has to be converted properly to depth. The conversion is based on the assumption that the sample material is removed by the sputtering process layer-by-layer to correlate accurately with the simultaneously recorded line intensities of the particular elements and the corresponding depth [163]. However, due to instrumental as well as sputter-induced effects this precondition is sometimes questionable in practice. As a result a sharp step-like change in the composition of a sample cannot be displayed with an adequate resolution, resulting in an apparently widened interface region [163]. An accurate calibration with properly chosen reference materials is required to achieve reliable depth-profiles of the sample composition [163].

The depth profile of the MMNC film composition was determined by Dr. Manfred Baumgärtner and Harald Merz (Research Institute Precious Metals & Metals Chemistry, Schwäbisch Gmünd) by means of GD-OES using a GDS 750 (Spectruma Analytik GmbH, Germany). The system was calibrated using standard reference materials, *i.e.* pure nickel and nickel alloys (IARM - Analytical Reference Materials International, USA). Finally, the raw intensity versus time data has been converted to elemental concentrations versus depth.

2.3.2 Transmission Electron Microscopy (TEM)

The transmission electron microscopy (TEM) investigations were performed in collaboration with Dr. Dagmar Dietrich and Dr. Thomas Lampke (Technische Universität Chemnitz, Institute of Composite Materials and Surface Technology). For TEM the composite films were separated from the substrate mechanically. Disks of 3 mm diameter were cut and ion polished by 3kV Ar ions with 6° incidence angle until electron transparency. The utilized TEM (Hitachi 8110) is equipped with a LaB_6 cathode and was operated at 200 kV which enables imaging in high resolution complemented by diffraction patterns, bright (BF) and dark field (DF) imaging.

2.3.3 X-ray Diffraction (XRD)

An often observed structural feature of polycrystalline electroplated films is that certain crystallographic lattice planes can occur with a greater probability than others, which is called preferred orientation or texture [159]. The occurrence of a texture can be easily recognized in a X-ray diffraction pattern by a pronounced enhancement of a certain reflection when compared with a powder pattern of randomly oriented grains. According to Scherrer, the width of the Bragg reflection increases with decreasing crystallite size [164]. However, it has to be emphasized that the crystallite size calculated from the XRD peak profiles refers to the domain sizes that scatter the incoming X-rays coherently. Hence, these values are generally smaller compared to the crystallite size as obtained by other techniques such as transmission electron microscopy [159].

X-ray diffraction was utilized to study the texture and crystallite size of the nickel matrix. Diffractograms were recorded by Ms. Anja Bensch (Inorganic Chemistry, Technische Universität Dresden) at room temperature on a Siemens D5000 with a scan rate of 0.12° min^{-1} for two-theta ranging from 10 to 100°. The size of the nickel crystallites was determined from the broadening of the (200) and/or (111) reflections according to the Debye–Scherrer equation [164], after background correction and subtracting the instrumental line broadening. The preferred crystalline orientation of the nickel films was evaluated by the relative texture coefficients RTC_{hkl}, Eq. (7)

$$RTC_{hkl} = \frac{I_{hkl}/I_{hkl}^0}{\sum_{1}^{4} I_{hkl}/I_{hkl}^0} \cdot 100\% \qquad (7)$$

where I_{hkl} are the relative intensities of the (hkl) lines measured in the diffractogram of the nickel films and I_{hkl}^0 are the corresponding intensities of a randomly oriented nickel powder sample (JCPDS no. 4-850). The summation $\sum I_{hkl}$ was taken for the 4 lines visible in the diffraction spectra, i.e. (111), (200), (311) and (222).

2.3.4 Vickers Microhardness

The mechanical properties of surface finishings are an important factor for their implementation in industrial processes [15]. The determination of the microhardness is often the method of choice for a straightforward screening because it is relatively inexpensive, easy to use and almost non-destructive.

A FischerScope HM2000 S was used for determination of Vickers microhardness of the coatings as described in DIN EN ISO 14577. The thickness of all coatings is more than ten times the maximum indentation depth of 1 μm in order to reduce effects of the substrates. Ten hardness tests were performed for each sample. The standard deviation was typically 10 %.

2.3.5 Abrasion Resistance

The bi-directional sliding abrasion resistance of the coatings was determined using a Taber Linear Abraser (Model 5750, Taber Industries) fitted with an Wearaser (CS 17, Taber Industries) consisting of alumina particles (particle size between 50 and 200 μm) in an elastic matrix. Taber linear abrasion testing is an industrially-accepted method to evaluate the abrasion resistance of surface coatings [165]. The tests were done under unlubricated conditions at room temperature and in ambient air. A load of 1950 g, a stroke distance of 1.27 cm, a sliding speed of 60 cycles per minute and a total number of 10,800 cycles was chosen for all the wear tests. After the test, the samples were sonicated in acetone for 5 minutes. The abrasion resistance of the coatings was characterized by weighting the samples with an accuracy of 0.1 mg before and after each abrasion test (each value is the average of at least three measurements). The weight loss of the samples, normalized by the mass of the analyzed abrasion volume, was taken as an indicator for the abrasion resistance of the coatings. A large relative weight loss indicates a low abrasion resistance.

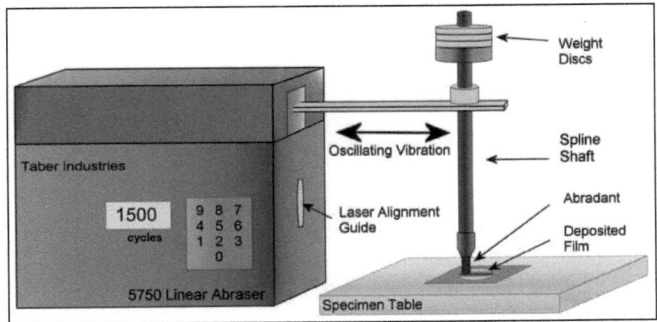

Figure 14 Schematic sketch of the measuring principle of the Taber Linear Abraser.

2.3.6 Magnetization Measurements

Vibrating Sample Magnetometry (VSM) is used to determine the magnetic properties of thin layers and small crystals of various nature (magnetic oxides, etc.) [141]. When a sample of any material is placed within a uniform magnetic field, a magnetic moment is induced in the sample. If this sample is made to undergo sinusoidal motion (*i.e.* mechanically vibrated), the vibration induces a magnetic flux change [141]. This in turn induces a voltage in the pick-up coils which is proportional to the frequency and amplitude of the sinusoidal motion, and the total magnetic moment of the sample at the particular applied magnetic field. In a VSM measurement the strength of the magnetic field is progressively changed while the magnetic moment of the sample is detected. Finally, hysteresis loops, *i.e.* plots of the magnetization (M) as a function of the applied magnetic field (H), are generated. Using these plots, the saturation magnetization (M_s), that represents the maximum magnetic moment measured, and the coercivity (H_c), that is the field required to reduce a saturated sample to zero induction, can be determined.

Hysteresis curves were measured by Dr. Christian P. Gräf (Saarland University, Department of Engineering Physics) at room temperature in a vibrating sample magnetometer (VSM, EG&G PARC 4500). In general the magnetization curve of a material depends on the composition and shape of the sample [166]. As the aspect ratio (diameter to layer thickness) of our magnetic layers is large no shearing correction is needed for hysteresis measurements with H_{mes} in the x-y-direction (Fig. 9, p. 28). Therefore all VSM measurements were performed in that direction.

3 Results and Discussion

3.1 Particle Characterization

3.1.1 Alumina

The Al_2O_3 Alu C particles exhibit a spherical or elliptical shape (Fig. 15). While the individual particles feature a size of 10-20 nm, the powder mainly consists of chain like agglomerates (Fig. 15). The 50 nm Al_2O_3 nanoparticles (Micropolish II) used for the IJE experiments (section 3.4) are spherically shaped with a size in the range of 30-60 nm [95].

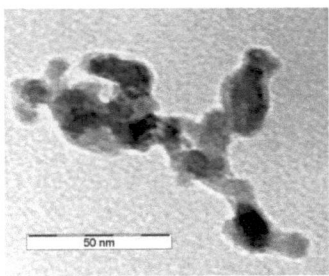

Figure 15 High-resolution bright field TEM image of the as-received 13 nm Al_2O_3 particles.

The pH dependence of the zeta (ζ) potential of 13 nm Al_2O_3 nanoparticles in 10^{-3} M KCl is shown in Fig. 16 (squares). From pH 2 to 6 the ζ-potential is almost constant and positive. At higher pH values the ζ-potential decreases, reaching the isoelectric point (IEP) at pH 9.2 [107]. In the absence of complexing agents, solutions containing Ni^{2+} tend to form nickel(II)hydroxide precipitates at higher pH values. Hence, the measurements involving nickel salts cover only the pH range 2-6. The addition of the sulfamate bath led to positive zeta potentials in the investigated pH range (Fig. 16, circles), probably due to the adsorption of nickel cations. In the case of the alkaline pyrophosphate bath the ζ-potential remains negative and no IEP could be determined within the investigated pH range (Fig. 16, triangle). This result can be explained by the adsorption of citrate or pyrophosphate anions [167]. Independent proof of the adsorption of particular ions comes from ζ-potential measurements in the presence of the individual bath components [16]. The working pH of the electrolytes is 4.3 for the sulfamate and 9.5 for the pyrophosphate bath (Table 1, p. 22). Hence, we can conclude that in the alkaline electrolyte the particles will be charged negatively and in the acidic sulfamate bath positively. Although, the absolute values of the ζ-potential will be smaller in the undiluted electrolytes due to the high ionic strength the general tendency will be the same [23, 168].

Using PCS (section 2.2.2), the particle size in the electrolyte was determined at the working pH of the two nickel plating baths. The mean hydrodynamic diameter of the Al_2O_3 particles (Alu C) was ~130 nm in the sulfamate electrolyte and ~70 nm in the pyrophosphate bath. The ECD process is obviously affected by the particle characteristics (section 1.4). Due to the differences in the alumina particle size in the two nickel electrolytes, one would expect distinct differences in the particle incorporation behavior, i.e. the amount and distribution of particles embedded in the nickel matrix [116].

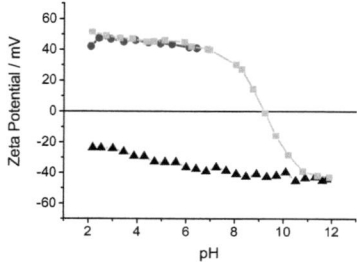

Figure 16 Zeta potential of 13 nm Al_2O_3 nanoparticles (0.2 g l^{-1}) in diluted 10^{-3} M electrolytes as a function of pH. (■) KCl; (●) sulfamate electrolyte; (▲) pyrophosphate electrolyte.

3.1.2 Titania

Figure 17 shows a TEM image of the as-received TiO_2 nanoparticles. It can be seen that they are spherical with a diameter between 15 and 30 nm. Hence, the diameter of 25 nm given by the producer can be considered as a mean value for the dry nanopowder. The particle size in an actual plating bath varies with the composition and the pH of the electrolyte (section 1.2.2). At the working pH of the two plating baths, a TiO_2 agglomerate size of about 430 nm for the acidic sulfamate and a size of about 110 nm for the alkaline pyrophosphate bath were determined.

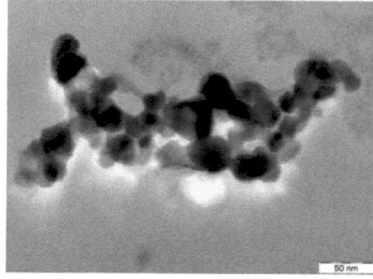

Figure 17 High-resolution bright field TEM image of the as-received TiO_2 particles.

The zeta (ζ) potential of the titania nanoparticles was investigated using the principle of micro-electrophoretic mobility (section 2.2.1). Figure 18 summarizes the pH dependence of the ζ-potential of TiO_2 nanoparticles in KCl as well as in the presence of the two diluted nickel baths. The ζ-potential of the TiO_2 particles in 10^{-3} M KCl decreased with increasing pH (Fig. 18, squares). In accordance with Ref. [107] the isoelectric point (IEP) of titania was found at a pH of ~5.8. As discussed in section 3.1.1, solutions containing Ni^{2+} tend to form nickel(II)hydroxide precipitates at higher pH values in the absence of complexing agents. Hence, the measurements involving nickel salts cover only the pH range 2-6. In the acidic sulfamate electrolyte the ζ-potential is positive in the pH range from 2 to 6 (Fig. 18, circles). In the presence of the alkaline pyrophosphate bath the pH-ζ curve is shifted to negative values and no IEP could be found in the pH range 2-12 (Fig. 18, up triangles). The working pH of the electrolytes are 4.3 (sulfamate) and 9.5 (pyrophosphate), respectively (Table 1, p. 22). The results shown in Fig. 18 indicate that the TiO_2 particles will bear a negative surface charge in the alkaline bath and a positive one in the acidic bath. Of course, the thickness of the EDL around the particles depends on the electrolyte concentration. Due to the high ionic strength of the plating baths the thickness of the diffuse part of the EDL will be reduced resulting in a change of the absolute value of the zeta potential. In general, the surface charge of oxide surfaces is a complex function of the pH and ionic strength of the solution [169]. However, the sign as well as the general tendency of the ζ-potential should be unaffected by the change of the electrolyte concentration [23, 168].

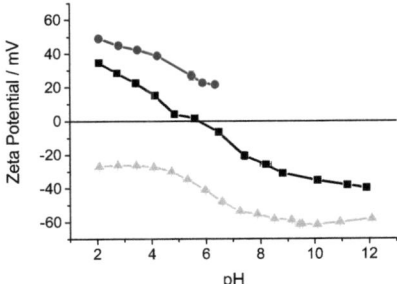

Figure 18 Zeta potential of TiO_2 nanoparticles (0.2 g l^{-1}) in diluted 10^{-3} M electrolytes as a function of pH. (■) KCl; (●) sulfamate electrolyte; (▲) pyrophosphate electrolyte.

To investigate the effect of the individual bath components on the ζ-potential of the TiO$_2$ particles we measured the ζ (pH) curve in the 10^{-3} M solutions of the corresponding components (Figs. 19a and b). The addition of citric acid and potassium pyrophosphate shifted the ζ-potential to negative values and the IEP to lower pH values (Fig. 19b). The presence of the nickel salts (nickel sulphate, nickel sulfamate, nickel chloride) leads to positive zeta potentials, probably due to the adsorption of nickel cations (Figs. 19a and b). Boric acid had almost no influence on the pH-ζ dependence and the position of the IEP of the titania particles (Fig. 19a). Thus, from the data shown in Figs. 18 and 19 we can conclude that the shift of the IEP in the nickel baths is caused by the adsorption of citrate and pyrophosphate anions (alkaline bath) and nickel cations (acidic bath), respectively.

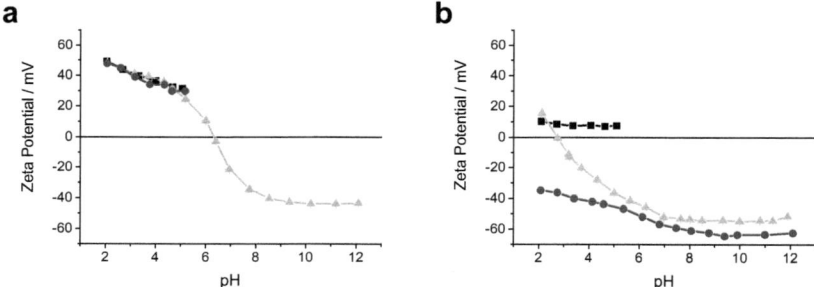

Figure 19 Zeta potential of TiO$_2$ nanoparticles (0.2 g l^{-1}) in different electrolytes as a function of pH (all concentrations are 10^{-3} M). (a): (■) nickel sulfamate; (●) nickel chloride; (▲) boric acid; (b): (■) nickel sulphate; (●) potassium pyrophosphate; (▲) citric acid.

The stability of the dispersed titania particles in the nickel electrolytes was investigated by the sedimentation method using a special centrifuge with an integrated optoelectronic sensor system (section 2.2.3). Figure 20 shows the sedimentation-time curves measured at a particle concentration of 5 g l^{-1} titania with a rotation speed of 500 rpm for the two nickel electrolytes. These curves represent the motion of the boundary between the clear phase (electrolyte) and the sediment toward the cuvette bottom as a function of time. The indicated radius is the distance of the boundary from the rotation axis (Fig. 13, p. 34). During the sedimentation process the position of the interphase shifts towards higher radial values. The different slopes in the sedimentation curves indicate that the nanoparticles sediment faster in the acidic sulfamate bath. This is in agreement with the different size of the agglomerates in the two nickel baths which were discussed above.

The sedimentation velocities calculated from the slopes of the sedimentation-time curves recorded at a rotation speed of 500 rpm are 350 μm s^{-1} in the acidic sulfamate bath and 220 μm s^{-1} in the alkaline pyrophosphate bath, respectively. These results are in accordance with the zeta potential of titania in the nickel electrolytes (Figs. 18 and 19). The adsorption of the components of the alkaline bath, citric acid and potassium pyrophosphate, induces a negative surface charge which leads to an improved electrostatic stabilization of the dispersion [16, 17].

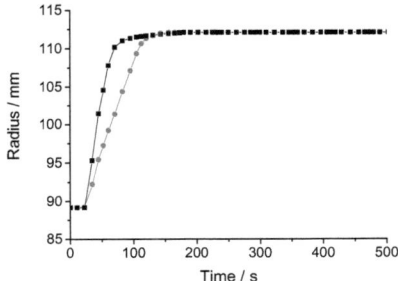

Figure 20   Sedimentation curves of TiO$_2$ nanoparticles (5 g l^{-1}) in different electrolytes at a rotation speed of 500 rpm. (■)sulfamate electrolyte; (●) pyrophosphate electrolyte.

3.1.3 Cobalt nanoparticles

As can be seen in the bright field TEM images (Fig. 21), the synthesized Co nanoparticles were either cubically or discoidly shaped. The characteristic edge length of the Co cubes was about 50 to 60 nm (Fig. 21a), whereas the Co discs have a diameter of about 15 to 50 nm (Fig. 21b). Close inspection of the micrograph (Fig. 21b) reveals some vertically aligned nanoparticles which proves that these are really discs.

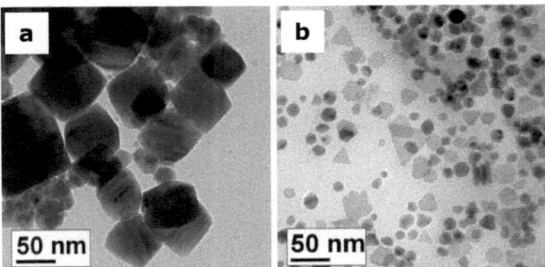

Figure 21   High-resolution bright field TEM image of the as-prepared Co nanoparticles. (a) cubes; (b) discoid.

The magnetic characterization and the XRD analysis of the Co nanoparticles have been discussed for cubes in Ref. [146] and discs in Refs. [170-172]. The magnetization measurements reveal the presence of a ferromagnetic/antiferromagnetic Co/CoO interface. The saturation magnetization at 295 K is about 139 A m^2 kg^{-1} which is about 85 % of the value for hcp Co single crystals (162 A m^2 kg^{-1}). The XRD data of the Co nanocubes indicate a mixture of hcp Co and ε-Co [173]. The XRD patterns for the Co nanodiscs indicate a hcp phase. The saturation magnetization of the discs is only 60% of the value for bulk Co.

3.1.4 Magnetite nanoparticles

In good agreement with Ref. [148], the magnetite nanoparticles were approximately spherical shaped with a narrow size distribution, as observed by high-resolution TEM microscopy (Fig. 22). The average particle diameter, estimated from the TEM micrographs was about 10-20 nm.

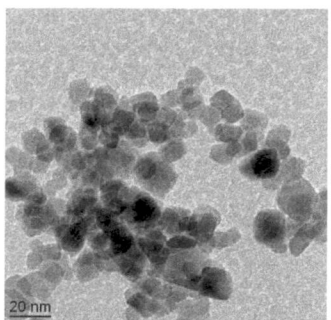

Figure 22 High-resolution bright field TEM image of the as-prepared Fe$_3$O$_4$ nanoparticles.

The XRD pattern demonstrates the crystalline structure and the composition of the synthesized magnetite powder (Fig. 23). The positions and relative intensities of all diffraction peaks coincide with those of the standard magnetite powder sample (JCPDS no. 19-0629). The crystallite size, calculated from the Scherrer equation (section 2.3.3) is 16 nm, which is in perfect agreement with the TEM data presented above.

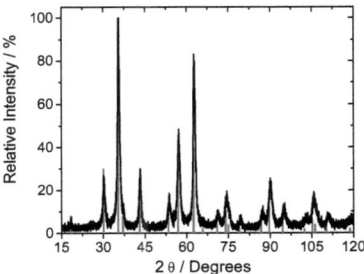

Figure 23 XRD pattern of Fe_3O_4 nanoparticles. (—) measured pattern; (—) reference pattern (JCPDS no. 19-0629).

The magnetite nanoparticles show a super-paramagnetic behavior as can be seen from the magnetization curves (Fig. 24) [174]. The saturation magnetization of the magnetite particles was about 65 A m^2 kg^{-1}. This value is obviously smaller than that of bulk magnetite, 92 A m^2 kg^{-1}, which is probably due to the small size of the particles [151, 175].

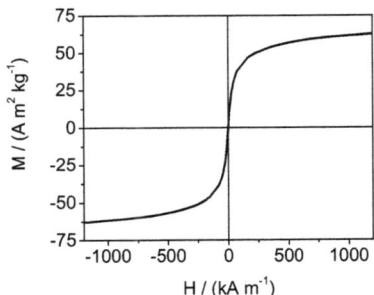

Figure 24 Hysteresis loop of the as-prepared Fe_3O_4 nanoparticles.

In the alkaline pyrophosphate electrolyte (Table 1, p. 22), the magnetite particles feature a negative surface charge in the pH range from 6.5 to 11.5 with an isoelectric point (IEP) of 6.5 (Fig. 25). The hydrodynamic diameter of the particles is directly related to their surface charge. In the pH range between 3-4.5 and 6.5-11.5, the hydrodynamic radius of the magnetite particles is below 100 nm. These relatively small diameters coincide with high values of the zeta potential (section 1.2.2). Between pH 4.5 and 6.5 where the zeta potential goes to zero a strong increase of the hydrodynamic diameter is observed which can be explained with agglomeration of the particles (Fig. 25). The working pH value of the Ni pyrophosphate electrolyte is 9.5 (Table 1, p. 22). Hence, the magnetite particles are charged negatively probably due to the adsorption of citrate and pyrophosphate anions (Fig. 19b, p. 44). The average hydrodynamic diameter of the particles is about 65 nm, which is about 3 times larger than the diameter observed in TEM (Fig. 22) and XRD measurements (Fig. 23). These differences are probably due to the formation of small aggregates in the electrolyte.

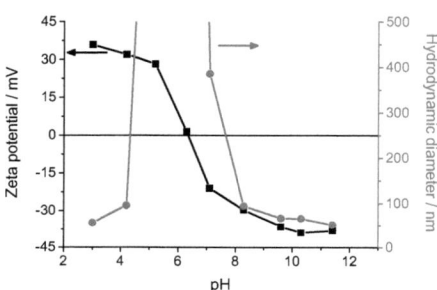

Figure 25   Zeta potential and hydrodynamic diameter of the Fe_3O_4 nanoparticles in the diluted nickel pyrophosphate electrolyte as a function of pH.

3.2 Alumina Particle Adsorption on the Nickel Electrode

In order to gain insight in the particle codeposition mechanism, the adsorption behavior of alumina nanoparticles on two kind of metal surfaces, *i.e.* Au and Ni, was studied as a function of the electrolyte flow rate and the electrode potential. This section focuses on the results obtained using a nickel electrode, since it is of major importance for the present work. A sophisticated EQCM has been used as a gravimetric sensor to quantify the attachment of small quantities of alumina particles on the electrode. For all EQCM measurements, the ratio between the frequency shift and the FHWM shift was above one, indicating that Eq. 6 (p. 29) can be used to calculate the mass adsorbed on the quartz.

Figure 26a shows the effect of electrolyte flow rate on the adsorption of 13 nm diameter Al_2O_3 particles on the nickel electrode. The flow rate was varied from 0.11 to 0.6 l min^{-1}. In general, there is a tendency of increasing particle adsorption with increasing electrolyte flow rate (Fig. 26a) which seems to be rather stronger in the range of low pH values. Due to a higher electrolyte flow rate more particles can arrive at the electrode which finally improves the particle adsorption reaction. Regardless of the particular electrolyte flow rate and electrode potential, a distinct increase in the mass adsorbed on the nickel electrode appeared with decreasing electrolyte pH from 9.5 to 4.3 (Fig. 26a).

Within an electrolyte of high ionic strength, the nanoparticles show a tendency to aggregate which has been proven in particle size and dispersion stability measurements (section 3.1). In general, the aggregation behavior can be influenced by different mechanisms, *e.g.* steric and/or electrostatic stabilization [101]. Electrostatic stabilization is due to the charge on the particle surface, which is defined by the composition and pH of the surrounding media as well as by the chemical composition of the particle surface (3.1). In an inert electrolyte (*e.g.* KCl) the alumina particles represent a positive surface charge in the pH range from 2 to 9.2 with an isoelectric point of 9.2 (Fig. 16, p. 42). An improved particle adsorption appeared at lower electrolyte pH (Fig. 26a) which in turn implies a higher positive surface charge on the alumina particles (Figs. 16). This improved particle adsorption on the electrode from an acidic electrolyte can be rationalized by comparison of the charges on the electrode as well as on the particle surface. Since the experiments shown in Fig. 26a took place at a constant cathodic electrode potential of -750 mV vs. Ag/AgCl, which is cathodic with respect to the potential of zero charge of a nickel electrode in a KCl electrolyte [176], the electrode will bear a negative charge. Hence, the results shown in Figs. 16 (p. 42) and 26 indicate that the surface charge of the alumina nanoparticles is directly related to the total mass adsorbed on the nickel electrode (Fig. 26a). In the pH range 2-9.2, the attractive interactions between negatively charged electrode and positively charged particles increase with the absolute value of the particle surface charge, *i.e.* with decreasing electrolyte pH.

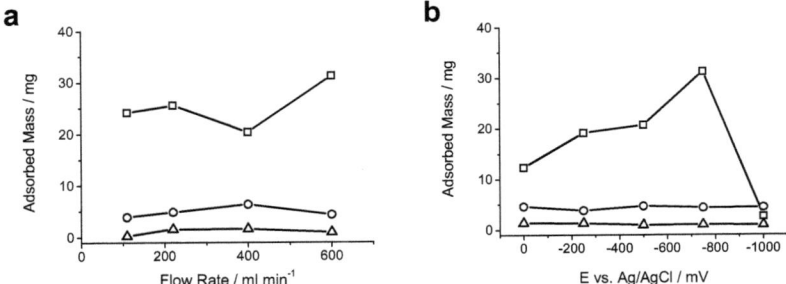

Figure 26 Mass adsorbed onto a freshly nickel coated quartz as a function of (a) electrolyte flow rate at an electrode potential of -750 mV vs. Ag/AgCl and (b) electrode potential at a flow rate of 600 ml min^{-1}. Electrolyte composition: 5 g l^{-1} Al$_2$O$_3$ in 1 M KCl. (□) pH 4.3; (○) pH 7.5; (Δ) pH 9.5.

The effect of electrode potential on the particle adsorption is given in Fig. 26b. The electrode potential was varied cathodically between 0 and -1000 mV vs. Ag/AgCl. While no change in the total mass adsorbed on the nickel electrode can be observed in the case of a neutral or alkaline electrolyte solution, a complex dependence appears at an electrolyte pH of 4.3 (Fig. 26b). The particle adsorption rate increases almost linearly to a maximum value at -750 mV vs. Ag/AgCl and then decreases. The beneficial effect of a more cathodic electrode potential on the particle adsorption might be due to an improved electrostatic interaction between positively charged particles and negatively charged electrode. It is evident that this effect appears preferentially at a low electrolyte pH (Fig. 26b), where the particles bear a distinct positive surface charge (Fig. 16, p. 42). Increasing the electrode potential from 0 to -1000 mV vs. Ag/AgCl leads to an obvious increase in current density (Fig. 27). Consequently, the drop of the mass adsorbed on the electrode in the range of high electrode potentials (Fig. 26b) can be explained by the intensification of side reactions, e.g. hydrogen evolution, which can either attenuate the particle transport to the electrode or cause a particle desorption. Another important side reaction is the oxygen reduction which might bring about a noticeable pH increase in front of the nickel electrode [119, 131]. An increase in the electrolyte pH leads to a decreasing absolute value of the surface charge of the Al_2O_3 particles (Fig. 16, p. 42) which finally affects the electrostatic interactions between particles and electrode.

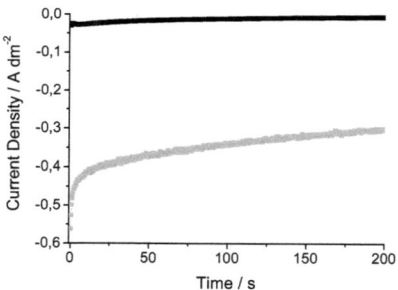

Figure 27 Current density transients during the EQCM adsorption measurements at a flow rate of 600 ml min^{-1}, a pH of 4.3 and an electrode potential of (—) 0 or (—) -1000 mV vs. Ag/AgCl. Electrolyte composition: 5 g l^{-1} Al_2O_3 in 1 M KCl.

A direct correlation of the recent particle adsorption experiments with the ECD particle incorporation results is challenging since the electrode potential is typically more cathodic during the ECD process. Moreover, no metal deposition reaction has been regarded so far in the adsorption experiments. Nevertheless, the obtained results are in good agreement with the ECD model of Guglielmi [49] and emphasize the significance of two particular processes, i.e. physical adsorption and electrophoresis on the process of particle adsorption on the metal electrode.

3.3 Electrocodeposition of Ni-Al$_2$O$_3$ and Ni-TiO$_2$ with the Parallel Plate Electrode

As extensively described in the literature the ECD process is affected by a variety of working conditions, such as current density, hydrodynamics, bath composition, pH and particle loading of the electrolyte [2, 3, 33, 50]. Hence, the first step of this investigation was to determine the effects of these parameters on the nickel deposition. Following, the effect of the process parameters on the particle incorporation was studied. The ECD of Ni-Al$_2$O$_3$ and Ni-TiO$_2$ nanocomposites was investigated using a parallel plate electrode configuration (section 3.3.1) and various current modulations (section 3.3.2). The results of DC plating are given in detail in Refs. [16, 177] and those of pulse plating (PP) and pulse reverse plating (PRP) in Ref. [117]. The results for the ECD of nickel alumina nanocomposites using the IJE system can be found in detail in Ref. [178] and are summarized in section 3.4. Based on the experimental results from the ECD of 50 nm alumina particles in nickel using an IJE system, a model was developed to predict the volume fraction of particles in a composite film by analyzing the particle flux to the electrode (section 4). While the experiments with the IJE system (3.4) have been performed using the 50 nm alumina particles (Micropolish II, Table 2, p. 23), the depositions with the PPE setup (3.3.1 and 3.3.2) were carried out using the 13 nm alumina particles (Alu C, Table 2, p. 23).

3.3.1 Direct Current Deposition

ECD of Ni-Al$_2$O$_3$ composite films

The relationship between the amount of co-deposited 13 nm Al$_2$O$_3$ particles in the composite coating and the particle loading of the electrolyte is shown in Fig. 28 for different plating current densities.

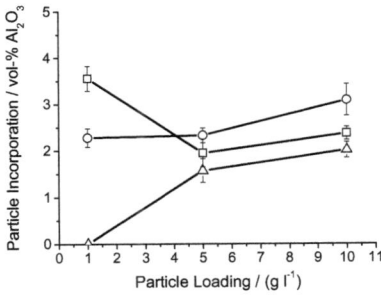

Figure 28 Correlation between the Al$_2$O$_3$ content in the layer and the particle loading of 13 nm Al$_2$O$_3$ particles. (□) 1; (○) 5; (Δ) 10 A dm^{-2}.

The particle incorporation in the nickel matrix ranged from about 0 to 3.6 vol-%. Low values of the current density were found to be beneficial for the ECD. A maximum incorporation of about 3.6 vol-% Al_2O_3 was obtained at a current density of 1 A dm^{-2} and a particle loading of 1 g l^{-1}.

ECD of Ni-TiO₂ composite films

The relationship between the amount of co-deposited TiO_2 nanoparticles in the nickel matrix and the plating conditions is shown in Fig. 29 for the different nickel plating baths (Table 1, p. 22). The amount of particle incorporation varied between 0 and 10.4 vol-% TiO_2 depending on the composition and pH of the electrolyte as well as the plating conditions. The maximum incorporation of about 10.4 vol-% TiO_2 was found in the case of the alkaline pyrophosphate bath at a current density of 1 A dm^{-2} and a particle content of 10 g l^{-1} TiO_2 (Fig. 29b). In comparison, using the acidic sulfamate electrolyte at the same plating conditions yielded nanocomposites containing only 7.9 vol-% TiO_2 (Fig. 29a).

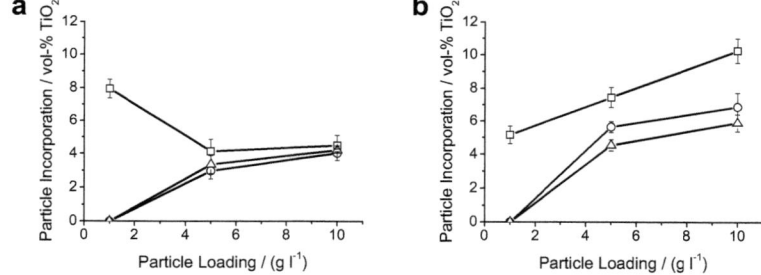

Figure 29 Correlation between the TiO_2 content in the layer and the concentration of TiO_2 in the electrolyte for different current densities. (□) 1; (○) 5; (Δ) 10 A dm^{-2}: (a) sulfamate electrolyte; (b) pyrophosphate electrolyte.

Both, the nanocomposites plated from the acidic sulfamate bath (Fig. 29a) as well as those from the alkaline pyrophosphate bath (Fig. 29b) exhibited an increase in the particle incorporation with increasing particle content of the electrolyte and decreasing plating current density. This behavior can be understood in terms of the two-step adsorption model of Guglielmi (section 1.3.1) which quantitatively describes the influence of particle content in the electrolyte and the current density on the particle incorporation [49]. In the first step the particles become loosely adsorbed on the electrode surface and stay in equilibrium with the particles in solution [179]. An increase of the particle concentration in the electrolyte causes an increase of the particle adsorption rate.

In the second step the shell of adsorbed ions is broken by the electrical field at the interface followed by a strong adsorption of the particles onto the electrode. The dependence of the particle deposition on the cathode potential is described by a Tafel expression [49]. It is interesting to note that the rate of particle incorporation is higher at low current densities and levels off at higher current densities (Fig. 29). This dependence of particle incorporation on the current density is consistent with Guglielmi's model [49]. At low current densities the ECD process is controlled by the particle adsorption and hence the particle incorporation is dominant. The asymptotic behavior of the particle incorporation with increasing current density can be attributed to the faster deposition of the metal matrix [86].

In agreement with previous results for the system $Cu-Al_2O_3$ and $Ni-Al_2O_3$ [16, 17], the films deposited from the alkaline bath contained significantly higher amounts of particles. Hence, there seems to be a tendency that negatively charged oxide particles are preferentially co-deposited in cathodic processes, at least in solutions containing divalent cations. To explain this at first glance counter-intuitive behavior an electrostatic model was proposed [16, 17] that takes into account the charge distribution on the particle and electrode surface. In the alkaline electrolyte the TiO_2 nanoparticles are negatively charged, whereas in the acidic one they bear a positive charge (Fig. 18, p. 43). Under the condition of the nickel electroplating process, the electrode bears negative excess charges [16, 17]. According to the model described in Refs. [16, 17], negatively charged particles are preferentially attracted by the positive excess charges in the electrolytic part of the electrical double layer of the electrode. When the particle has come close to the electrode the shell of adsorbed ions on the particle is stripped off within the EDL of the electrode. Finally the particle becomes incorporated into the growing metal layer.

3.3.2 Pulse Plating and Pulse Reverse Plating

Based on the results of direct current (DC) plating, the PP and PRP conditions were optimized in order to create compact nickel surface finishings with a smooth surface morphology. The main results concerning the influence of the current modulation on the codeposition of $Ni-Al_2O_3$ nanocomposites are summarized as follows.

Pulse Plating (PP)

The process parameters investigated with the PPE system were current density of 1-10 A dm^{-2} and particle loading of up to 10 g l^{-1}. PP codeposition of alumina particles with nickel was carried out using rectangular pulses with pulse frequencies ranging from 1.7 to 8 Hz, pulse duty cycles between 11 and 67 % and a cathodic peak current density of either 5 or 10 A dm^{-2} (Table 3, p. 26) [117]. The variation of the amount of co-deposited 13 nm alumina particles in the nickel matrix with the duty cycle and the pulse frequency is shown in Fig. 30 for various particle contents in the bath. The particle incorporation ranged from ~2.1 to ~6.2 vol-% alumina depending on the PP conditions. In general, particle codeposition increased with the particle loading in the electrolyte, the pulse frequency and with decreasing pulse duty cycle (Fig. 30) [117].

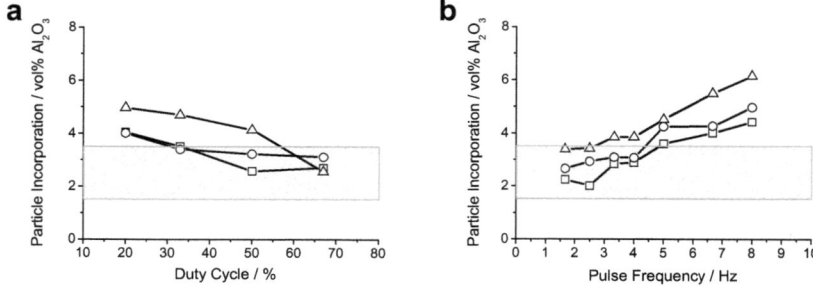

Figure 30   Effect of duty cycle (a) and pulse frequency (b) on the alumina content of Ni-Al_2O_3 composites. Particle loading: (□) 1; (○) 5; (Δ) 10 g l^{-1} of 13 nm Al_2O_3 particles. The range of incorporation data using DC deposition at various current densities and particle loadings is indicated by the hatched areas [16].

The effects of duty cycle and pulse frequency can be explained by the relationship of t_{on} and t_{off} (section 2.1.3.1). Decreasing the duty cycle as well as increasing the pulse frequency corresponds to a longer relaxation time, *i.e.* the time at which no current passes through the system. Hence, due to the longer pause more alumina particles arrive at the electrode without a "concurrent" metal deposition taking place in parallel, and as a result the particle incorporation increases. Compared to previous results on the ECD of Ni-Al_2O_3 composites using DC plating (Fig. 30, hatched area) [16], the incorporation of alumina nanoparticles in nickel significantly increased in the PP process, particularly at low duty cycles. The enhanced particle incorporation during PP can be understood in terms of the model of Tóth-Kádár et al. which has been developed to describe the varying microstructure of nickel during PP, by means of the metal ion concentration in front of the electrode ($c_{i,s}$) [180]. According to their model, the microstructure of PP nickel is similar to that of DC nickel when $t_{off} < t_{on}$ (*i.e.* θ > 50 %), because $c_{i,s}$ during PP will converge to $c_{i,s}$ of DC deposition at least after a certain number of PP cycles. However, the nickel microstructure is markedly different when $t_{off} > t_{on}$ (*i.e.* θ < 50 %) because $c_{i,s}$ is expected to recover to the value of the bulk electrolyte ($c_{i,b}$) due to the longer pulse pause. Hence, with regard to the ECD process the particle concentration in front of the electrode ($c_{p,s}$) might vary depending on the t_{on}/t_{off} ratio, *i.e.* the PP duty cycle θ. In the case of a low θ, $c_{p,s}$ is assumed to be similar to $c_{p,s}$ of DC deposition leading to a similar amount of particle incorporation (Fig. 30a). However at low θ, $c_{p,s}$ is assumed to recover to the bulk concentration ($c_{p,b}$) which finally leads to a higher particle content of the MMNC films (Fig. 30a).

The particle content of the nickel nanocomposites increased when the peak current density was decreased from 10 and 5 A dm^{-2} (Fig. not shown) [117]. This increase can be explained by the model of Celis et al. (section 1.3.2) which describes the particle codeposition as a multi-step process [1]. According to this model, the particle incorporation is mainly governed by the reduction of metal ions adsorbed on the particle surface. During PP deposition, the reduction of free metal ions as well as ions adsorbed on the particles takes place during t_{on}. The rate of metal reduction increases with the absolute value of the cathodic peak current density. Compared to ions adsorbed on the particles free metal ions in the solution have a higher mobility. The increasing peak current density leads to a preferential reduction of free ions at the electrode [75]. Hence, the particle content of the electrocodeposited composite films decreases with increasing peak current density. Independent proof of the adsorption of the metal cations comes from measurements of the zeta potential of alumina particles in sulphate containing solutions [16].

The effect of the individual pulse plating parameters (particularly pulse length, t_{on}, pulse pause, t_{off}, and peak current density, i_p) can be summarized by introducing the average plating current density, Eq. 8.

$$i_{ave} = \frac{t_{on} \cdot i_p}{t_{on} + t_{off}} = \theta \cdot i_p \qquad (8)$$

Taking into account that the average current density increases with the duty cycle, θ, and peak current density, i_p, one can conclude that the incorporation of alumina nanoparticles is favored by low average current densities.

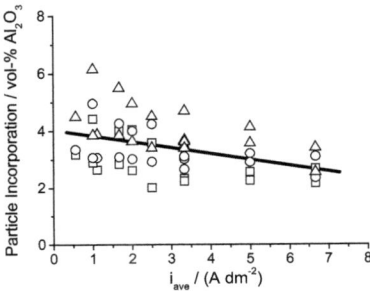

Figure 31 Effect of the average current density during PP on the particle incorporation in nickel for different particle loadings. (□) 1; (○) 5; (Δ) 10 g l^{-1} of 13 nm Al$_2$O$_3$ particles.

Figure 31 illustrates the particle content of the composites prepared by PP from electrolytes with loadings ranging from 1 to 10 g l^{-1} of 13 nm γ-Al_2O_3 particles. Particle incorporation showed a tendency to increase with decreasing average plating current density. A maximum incorporation of about 6.2 vol-% Al_2O_3 was found at an average current density of ~1 A dm^{-2} and a particle loading of 10 g l^{-1}. Similar to the described behavior, the highest incorporation in the case of DC plating of 3.6 vol-% Al_2O_3 in nickel was found at a relatively low current density of 1 A dm^{-2} [16].

Pulse Reverse Plating (PRP)

As a preliminary test of the PRP experiments, polarization curves of a freshly prepared composite layer in the acidic sulfamate bath were measured using a platinum RDE [117]. Particularly, the anodic branch was significantly affected by the presence of alumina nanoparticles; a shift of the anodic dissolution peak to more cathodic potentials and an obvious increase of the anodic dissolution current density were observed [74, 97, 117]. The magnitude of both, the shift of the peak as well as the increased anodic current density strongly depends on the alumina concentration in the electrolyte and on the rotation rate of the electrode. Thus, due to an increasing particle loading from 1 g l^{-1} to 5 g l^{-1} Al_2O_3 the maximum stripping current was found to increase from ~0.8 A dm^{-2} to ~4.9 A dm^{-2} [117].

The particle content of the nickel based composites plated by PRP is shown in Fig. 32 for different cathodic pulse lengths between 50 and 400 ms. For comparison, the average particle incorporation during DC plating is indicated in the figure by the hatched area [16]. The incorporation of 13 nm Al_2O_3 particles is obviously enhanced by longer cathodic pulse times as well as lower anodic peak current densities (Fig. 32).

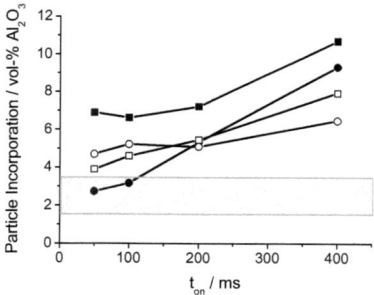

Figure 32 Effect of cathodic pulse time during PRP on the particle content of Ni-Al_2O_3 composites prepared at t_{rev}=20 ms from electrolytes containing 10 g l^{-1} of 13 nm Al_2O_3 particles. i_p=5 A dm^{-2}: (■) i_{an}=1 A dm^{-2}; (●) i_{an}=5 A dm^{-2}; i_p=10 A dm^{-2}: (□) i_{an}=1 A dm^{-2}; (○) i_{an}=5 A dm^{-2}. The range of incorporation data using PPE configuration at various current densities and particle loadings of 13 nm Al_2O_3 particles is indicated by the hatched area [16].

The maximum particle incorporation during PRP, *i.e.* 10.7 vol-% Al_2O_3 in nickel, was found at a cathodic pulse length of 400 ms, a cathodic peak current density 5 A dm^{-2}, an anodic pulse length of 20 ms and an anodic peak current density ~1 A dm^{-2} [117]. It has been reported in Refs. [54, 77] that the maximum particle incorporation during PRP was found when the metal deposit thickness approaches the particle diameter size. The effective diameter of the alumina nanoparticles in the nickel sulfamate electrolyte was determined to be ~130 nm (section 3.1.1). At a cathodic peak current density of 5 A dm^{-2}, 400 ms pulse length correlates to 7 nm layer thickness. Consequently, only a small fraction of the particle volume is incorporated during one cathodic pulse. Hence, it might be possible to further increase the amount of alumina incorporation by increasing the metal layer thickness per cathodic cycle. However, increasing both i_p as well as t_{on} resulted in deposits with a powdery appearance which did not adhere well to the substrate.

3.4 Electrocodeposition of Ni-Al$_2$O$_3$ with the Impinging Jet Electrode

The process parameters studied with the IJE system (section 2.1.3.2) were flow rate, current density and particle loading of the electrolyte. In contrast to the ECD experiments described so far, 50 nm Al_2O_3 particles have been used with the IJE system (Table 2, p. 23). The maximum possible bath content was 120 g l^{-1}, as sedimentation occurred quickly at higher loadings. Significant particle incorporation of ~1.6-7.7 vol-% was found with particle loadings as low as 10 g l^{-1}. Regardless of all the other plating parameters, maximum particle incorporation was found at flow rates of 2-3 l min^{-1} (Fig. 33). These results are quite similar to those reported by Osborne et al. for the codeposition of 50 nm γ-Al$_2$O$_3$ particles in copper [138]. Increasing the flow rate increases the quantity of particles arriving at the electrode, but reduces the residence time of particles at the electrode surface. The interaction between these effects influences the ECD process. As a result a decrease of the residence time of the particles at the electrode leads to an inhibition of particle codeposition. Interestingly the particle incorporation increases at a flow rate of 7.5 l min^{-1}. Higher levels of shear force acting on the electrode surface lead to a decrease in particle incorporation [55, 58, 60]. However, within the investigated flow region shear force might not be the most important factor in determining particle codeposition. Instead, the decrease in particle incorporation might be a result of the interaction between flow rate and limiting current density [63, 121]. Further investigations are needed with a higher capacity pump in order to determine the fluid flow range where hydrodynamic shear force effectively starts to affect the particle incorporation.

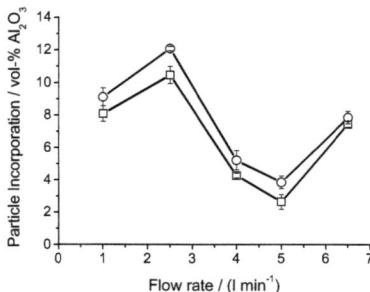

Figure 33 Correlation between the Al_2O_3 content in the layer and the flow rate for different particle loadings of the electrolyte at a current density of 10 A dm^{-2}. (□) 90, (○) 120 g l^{-1} 50 nm Al_2O_3 particles.

In Fig. 34 the effect of particle loading of 10 -120 g l^{-1} Al_2O_3 on the ECD with nickel is given at a current density of 10 A dm^{-2} for three different electrolyte flow rates. As expected, an increase in incorporation of particles with an increase in the loading in suspension was found [2].

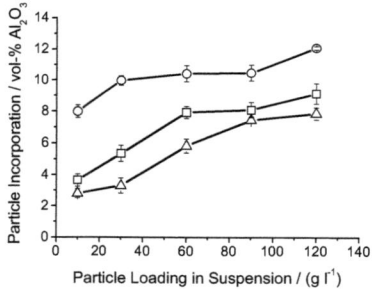

Figure 34 Effect of particle loading of the electrolyte on the particle incorporation in nickel at a current density of 10 A dm^{-2}. (□) 1; (○) 2.5; (Δ) 6.5 l min^{-1}.

To determine the effect of the current density on the particle incorporation, depositions were performed at three different flow rates with alumina concentrations of 10-120 g l^{-1} in the electrolyte as the current density was varied from 5 to 30 A dm^{-2}. The particle contents of the coatings plated with suspension loading of 90 g l^{-1} are shown in Fig. 35. As the current density was increased from 5 to 10 A dm^{-2} the particle incorporation increased significantly to a maximum value between 10-15 A dm^{-2} and then decreased for all flow rates. This behavior has been observed previously with both RDE and RCE systems [58, 60, 85], and is related to the balance of particle adsorption or entrapment versus metal deposition rate. A similar dependence of the particle incorporation on the plating current density appeared at all other particle contents of the electrolyte [181].

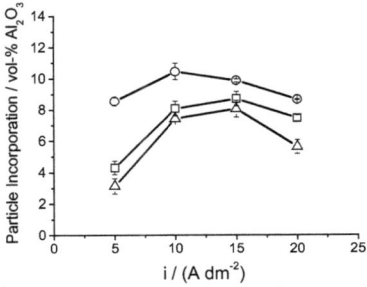

Figure 35   Influence of current density on the incorporation of alumina particles in nickel at particle loading of 90 g l^{-1} Al$_2$O$_3$. (□) 1; (○) 2.5; (Δ) 6.5 l min^{-1}.

The influence of hydrodynamics and current density can be summarized in a single parameter by normalizing the current density with the limiting current density [58]. The limiting current density of the IJE system was calculated according to the equation of Chin et al. [121, 178], which will be discussed in more detail in section 4.1 (Eq. 21, p. 87). As the percentage of limiting current density increased from about 5 to 10 %, the particle incorporation increased to a maximum value of 10.1 vol-% (Fig. 36). A further increase of the current density leads to a decrease in particle codeposition and finally levels off at a relatively constant value of particle incorporation (Fig. 36).

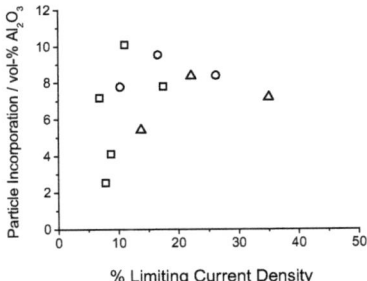

Figure 36   Particle incorporation vs. percentage of limiting current density at a particle loading of 90 g l^{-1} of 50 nm Al$_2$O$_3$ particles. (□) 10; (○) 15; (Δ) 20 A dm^{-2}.

The particle content of the composite films plated with the IJE was determined using the electrogravimetric method (section 2.3.1.1). The advantage compared to other methods, such as EDX, X-ray fluorescence, atomic absorption spectrometry, etc., is the fast and inexpensive determination of the particle content of the electroplated composites. As a validation of the incorporation results computed by the electrogravimetric method, EDX measurements were done on the surface as well as in the cross section of the composite films (section 2.3.1.2). Figure 37 compares the incorporation results obtained by the two methods. It is apparent that the contents of the EDX analysis are significantly higher than those of the electrogravimetric method. The differences in the incorporation results can be most probably attributed to differences in the size of the analyzed area [178]. While the area of analysis is limited to a few hundred µm² in the case of EDX (section 2.3.1.2), the electrogravimetric method accounts for the total volume of the composite film. Furthermore, a possibly uneven particle distribution within the composite film has to be considered, as it has been stated in the literature that particularly in the first micrometer of the coating exclusively metal is deposited [182, 183]. Also, in the case of the ECD of Ni-Al$_2$O$_3$ composites, the current efficiency of nickel deposition has been found to decrease considerably (from 80-90% to about 40%) due to the presence of the nanoparticles in the plating bath, particularly in the range where the particle incorporation is largest [97]. Hence, according to the procedure of the electrogravimetric analysis (section 2.3.1.1) [58], the overall lower particle contents determined by this method (Fig. 37) might also be caused by the effect of the alumina nanoparticles on the current efficiency of the nickel deposition [97].

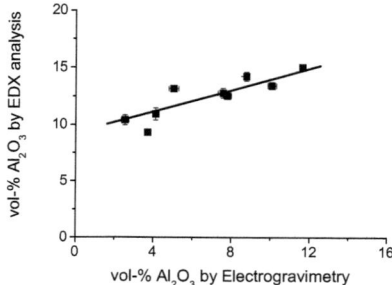

Figure 37 Particle content of the Ni-Al$_2$O$_3$ nanocomposites determined by EDX analysis as a function of the incorporation results of the electrogravimetric method.

Apart from the EDX analysis (Fig. 37), Al_2O_3 particle incorporation was validated by GD-OES (section 2.3.1.3). Figure 38 compares the amount of particles in the nickel layer obtained by these two techniques.

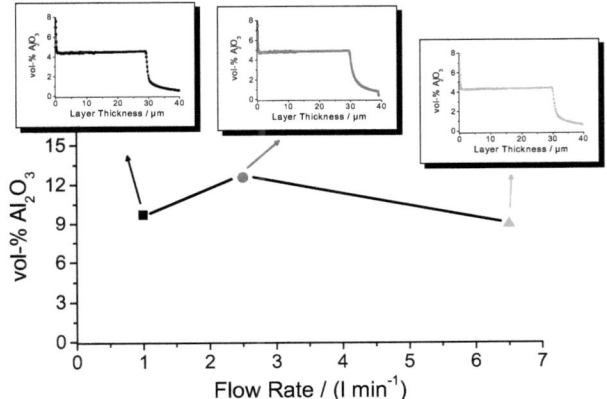

Figure 38 Al_2O_3 content (determined by means of EDX analysis) in a nickel film produced via Jet-Plating at a current density of 10 A dm^{-2} from an acidic sulfamate electrolyte containing 90 g l^{-1} Al_2O_3 nanoparticles. (■) 1; (●) 2.5; (▲) 6.5 l min^{-1}. The insets indicate the depth profile of the nanoparticle distribution in the film obtained by GD-OES analyses.

Although the GD-OES incorporation results are obviously smaller than those determined by EDX analysis, the overall trend with a change in the flow rate is similar, i.e. the highest amount of particles in the layer was found at a flow rate of 2.5 l min^{-1}. Additionally, the GD-OES analyses emphasize an overall uniform particle distribution within the metal matrix (Fig. 38). The differences of both methods, EDX and GD-OES, can be rationalized by the fact that a quantitative GD-OES analysis of the chemical film composition is based on a calibration with an appropriate reference material. However, since it is a challenging task to obtain composite layers with a given composition and/or particle distribution, the calibration of the GD-OES was achieved using pure nickel and nickel alloys as a reference material. In general, a homogenous particle distribution is necessary in order to obtain composite films with optimized properties [111]. Moreover, comparison of Figs. 37 and 38 emphasizes that the GD-OES incorporation results agree well with those of the electrogravimetric analysis. Compared to EDX analysis, both GD-OES and electrogravimetric analysis account for a bigger volume of the composite film. Hence, the overall lower incorporation results obtained by EDX analysis compared to GD-OES as well as the electrogravimetric method can be explained by the differences in the size of the analyzed area.

3.5 Electrocodeposition of Nickel Matrix Nanocomposites in a Magnetic Field

The relationship between the amount of co-deposited Co nanoparticles in the composite coating and the applied current density at a particle concentration of 6 g l^{-1} in the electrolyte is shown in Table 4 for disc-like and cubically shaped nanoparticles. The amount of co-deposited Co particles in the Ni matrix was approx. 20-40 vol-%. Compared to ceramic nanoparticles in Ni matrices (sections 3.3 and 3.4) [2, 16] these amounts of incorporation are up to ten times higher. This is probably due to the chemical similarity of Co and Ni. In general, it can be noted that the particle incorporation significantly increases with decreasing plating current density. A maximum incorporation of about 41 vol-% Co was obtained in the case of discoid particles (Table 4) at a current density of 5 A dm^{-2} in the presence of a 100 mT perpendicular magnetic field. Decreasing the particle content of the electrolyte from 6 to 3 g l^{-1} resulted in a decrease of the particle incorporation by 2 to 5 vol-% Co (data not shown). This would indicate that the particle incorporation is mainly governed by the chemical similarity between the Co nanoparticles and the Ni matrix, whereas a minor effect arises from the particle concentration in the electrolyte.

Table 4 Particle content of the Ni-Co composites prepared from electrolytes containing 6 g l^{-1} Co particles of either disk-like or cubically shape. **B** refers to the direction of the 100 mT magnetic field during electrodeposition: 0 no field; x - parallel; z - perpendicular to the electrode surface (see Fig. 9, p. 28).

Particles	B	i [A dm^{-2}]		
		1	5	10
cubes	0	34	30	26
	z	35	35	29
	x	/	24	/
discs	z	/	41	36
	x	/	25	/

Depending on the strength and the direction of the magnetic field the amount of Fe$_3$O$_4$ nanoparticles in the Ni matrix was 0 to ~4 vol-% (Fig. 39). Regardless of the presence of a magnetic field the magnetite incorporation was maximal between 0.5 and 5 A dm^{-2}. When the current density was further increased to 10 A dm^{-2} no Fe$_3$O$_4$ was co-deposited no matter what the flux density or orientation of the magnetic field was (Fig. 39). The leveling off of the particle incorporation at high current densities is a general tendency which has been observed for numerous systems [2, 3].

In the following, the effect of the 100 mT static magnetic field on the particle incorporation in different orientations, viz. parallel and perpendicular will be discussed. There will be a field gradient near the electrode surface due to the growing Ni layer that affects particle translation and incorporation (as well as alignment for the anisotropic Co particles). For the parallel orientation of **B** the particle incorporation showed a tendency to decrease (Table 4, Fig. 39). According to Eq. 4 (p. 18), a Lorentz force is generated under these conditions. For the orientation shown in Fig. 9a (p. 28) this volume force density, f_L, is directed parallel upwards to the working electrode and is supposed to drive convection in the same direction (magnetohydrodynamic or MHD convection) [128]. In the following the order of magnitudes of the forces which are involved in the ECD process will be estimated. These are the driving force of diffusion, $f_{diff} = k \cdot T \cdot \vec{\nabla} c$ (k Boltzmann constant, T absolute temperature, c concentration), the Lorentz force (Eq. 4, p. 18), and the magnetophoretic force (Eq. 5, p. 18). For this rough estimation migration forces have been neglected as we are dealing with media of high ionic strengths. Furthermore, the effects arising from the nitrogen bubbling have been neglected as this occurs relatively far from the electrode and we are only concerned with forces right in front of the electrode, say several hundred μm. A particle size of 100 nm and a magnetic susceptibility of 0.03 was used which is in agreement with the data shown in section 3.1.4. In the case of a parallel 100 mT magnetic field and a current density of 5 A dm^{-2} the Lorentz force acting on one particle is about 10^{-20} N. The force due to the concentration gradient is in the same order of magnitude (assuming a diffusion layer thickness of 100 μm). For the estimation of the magnetophoretic force the gradient of the magnetic field near the electrode is needed. Finite element simulations indicated that this value is in the order of 10 T m^{-1} for our thin Ni films and an external flux density of 100 mT. Hence, one arrives at an estimate of 10^{-17} N which means that the magnetophoretic force will be dominating the particle incorporation.

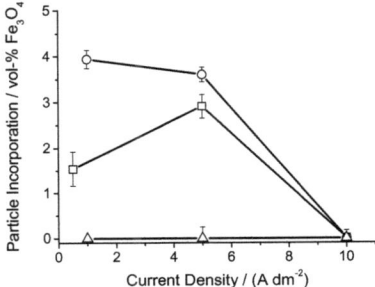

Figure 39 Comparison of the particle content of Ni-Fe$_3$O$_4$ composites plated at different current densities from electrolytes containing 10 g l^{-1} Fe$_3$O$_4$ nanoparticles. (□) **B** = 0 mT; (○) **B** = 100 mT, parallel (x-direction); (Δ) **B** = 100 mT, perpendicular (z-direction, see Fig. 9, p. 28).

Regardless of the particle type and shape, an increase of incorporation was found in the case of a perpendicular magnetic field (Table 4, Fig. 39). In a perpendicular magnetic field only weak magnetohydrodynamic electrolyte motion should occur because there is essentially no Lorentz force (Fig. 9b, p. 28; the cross product in Eq. 4, p. 18, yields zero). Note, that in a real electrochemical cell there will be always components of the current which are perpendicular to B (*e.g.* at the edges of the electrode) and therefore small contributions from the Lorentz force. However, the main effect should arise from the magnetic nanoparticles moving along the gradient of the B-field which is directed towards the working electrode as indicated by the trajectory f_{mp} in Fig. 9b. As a result, the particle concentration in front of the electrode is expected to increase, leading to enhanced particle incorporation. These considerations are in perfect agreement with the above estimations of the order of magnitudes of the forces and the experimental particle incorporation data (Table 4, Fig. 39). When B is applied in the z-direction (Fig. 9b, p. 28), increased particle incorporation was observed, provided the current density is not too high.

At a first glance one could expect that this magnetophoretic particle transport should also be active for the parallel B-field (*i.e.* aligned in x-direction, Fig. 9a, p. 28) because due to the presence of the magnetic field the growing Ni layer should be magnetized and thus attract the magnetic nanoparticles. However, the data in Table 4 and Fig. 39 clearly show that for this orientation fewer particles are co-deposited, in the case of magnetite none at all (Fig. 39). For this case the estimation of f_{mp} becomes complicated due to term $(B \nabla) B$ which involves the external and local B fields. Another contribution comes from the perpendicular electrolyte flow that is induced by the Lorentz force and carries particles away from the electrode (dotted lines denoted "MHD" in Fig. 9a, p. 28). Additional finite element simulations of the electrochemical cell will be necessary to estimate the trajectories of the nanoparticles near the surface for all orientations of B and j.

3.6 Structure and Properties

3.6.1 Surface Morphology

In the case of nickel electrodeposition from an acidic sulfamate bath, a bimodal surface structure of "truncated pyramidal type" was observed (Fig. 40a), where a cluster of fine grains surrounds relatively large grains [16, 117, 136, 178]. On the other hand, the pure nickel films from the alkaline bath are characterized by a smooth and fine grained surface structure over the investigated range of plating conditions (Fig. 40b). Due to the particle incorporation and the increase of the current density, a refinement of the surface structure appeared (micrographs not shown).

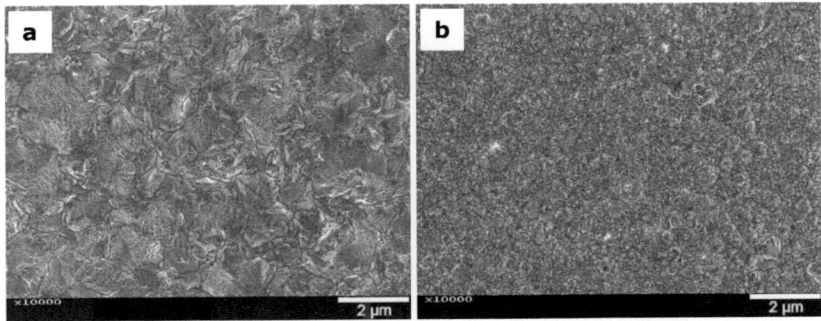

Figure 40 SEM surface morphology of pure Ni films plated at 5 A dm^{-2}; (a) sulfamate bath; (b) pyrophosphate bath.

The surface morphology of the electrodeposited nickel films, *i.e.* size and shape of the grains were significantly altered due to both, PP conditions and incorporation of Al_2O_3 and TiO_2 nanoparticles (Fig. 41). According to the electrocrystallization theory as discussed in detail by Budevski, Staikov and Lorenz, the grain size and growth mode is defined by the interplay of nucleation and growth [68]. In the case of PP and PRP experiments the number of atoms deposited during one cathodic cycle is defined by the pulse time, t_{on}, and the peak current density, i_p. Furthermore, sorption processes occurring during pulse pause may influence nucleation and growth processes in the following cathodic pulse cycle [68]. In our PP and PRP experiments (Tab. 3, p. 26), the cathodic pulse current density is comparable to the average current density (Eq. 8, p. 55) in DC plating. Finally, the cathodic overpotential during PP and PRP is close to that during DC plating. Hence, the surface morphology of the film is expected to be similar for the different deposition techniques, *i.e.* DC, PP and PRP (Figs. 40 and 41). The PP coatings showed a smooth surface with fine grains (Figs. 41a and c). The pure nickel films plated at a higher pulse current density and longer pulse on-time exhibited a coarser grain structure (Fig. 41a).

Higher values of the peak current density and cathodic pulse length correlate with an increased average current density (Eq. 8, p. 55). Hence, the coarsening of the nickel structure with increasing current density is in good agreement with the results for nickel plated from an acidic sulfamate bath under DC conditions [16, 184].

In the case of Ni-Al$_2$O$_3$ composites (Figs. 41b and d) the agglomerates of alumina particles appear as bright spots in the dark nickel matrix, as validated by EDX spectroscopy. The nickel growth is affected by several interfacial inhibitors such as H$_2$, H$_{ads}$, Ni(OH)$_2$ [70, 183, 185]. It is assumed that the particle incorporation perturbs the nickel growth and induces an increase of the number of nucleation sites resulting in a refined grain structure [3, 79, 186, 187]. Thereby the perturbation of the nickel growth results from a change in the adsorption-desorption phenomena at the nickel/electrolyte interface which is probably caused by a local pH increase induced by an adsorption of H$^+$ on the dispersed particles [76, 185].

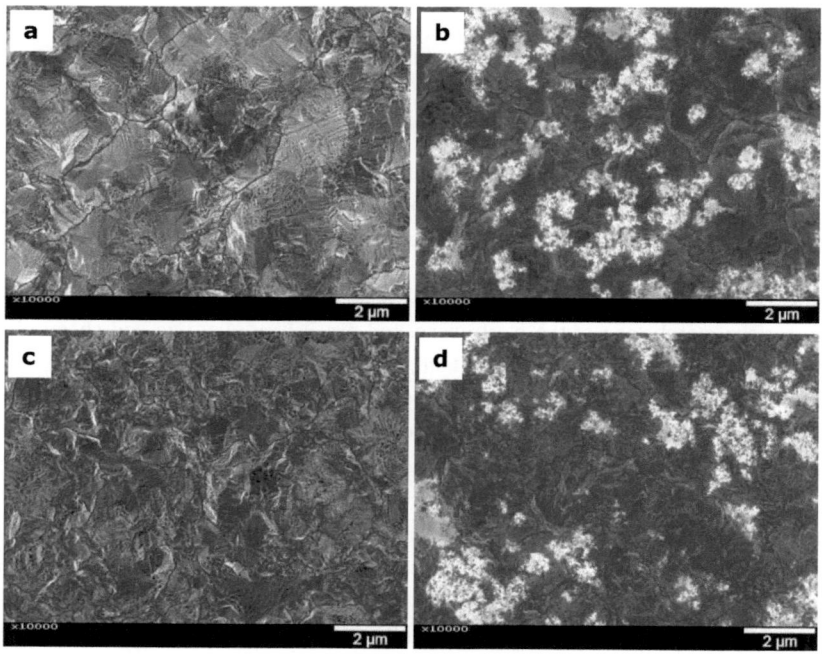

Figure 41 Surface morphology of Ni and Ni-Al$_2$O$_3$ composites plated at θ = 67 %, f = 1.67 Hz. 0 g l^{-1} Al$_2$O$_3$: (a) 10 A dm^{-2}, (c) 5 A dm^{-2}. 10 g l^{-1} Al$_2$O$_3$: (b) 10 A dm^{-2}, (d) 5 A dm^{-2}.

The pure nickel as well as the composite films deposited in the presence of a magnetic field (section 2.1.3.3) exhibited a smooth and fine grained surface morphology (micrographs not shown). In general, a refinement of the Ni grains appeared with increasing current density and increasing particle incorporation. The microstructure of the pure Ni films was not significantly affected by the application of a magnetic field during the deposition. A magnetic flux density of 100 mT is probably too low to induce strong changes. Moreover, our group noted that higher fields can have a pronounced effect on the surface morphology and the microstructure of metal depositions [188-191].

3.6.2 Microstructure

The pure nickel films plated from the acidic sulfamate bath exhibited a well-defined columnar structure belonging to the FT (field-oriented texture) type [16, 192]. Figure 42 compares the cross sectional morphologies of pure nickel films plated from the sulfamate electrolyte by means of three different techniques, *i.e.* DC, PP and IJE, respectively. It is apparent that the field oriented texture of the nickel films appeared regardless of the deposition technique, *i.e.* the particular electrode orientation and current modulation. Nevertheless, the quality of the columnar structure was affected by the particular plating conditions. In the case of PP deposition, an increase of the average plating current density (Eq. 8, p. 55) induced an obvious coarsening of the field-oriented nickel structure (micrographs not shown), which is directly related to the coarsening of the surface morphology (Figs. 41a and c, p. 66) [140].

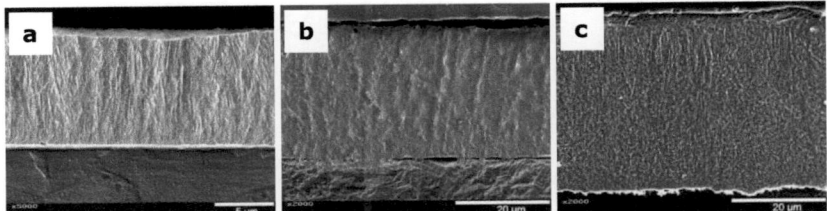

Figure 42 Cross-sectional SEM images of pure nickel films. (a) DC i =10 A dm^{-2}, (b) PP θ = 67 %, f = 1.67 Hz, i_p = 10 A dm^{-2}, (c) IJE 6.5 l min^{-1}, i =10 A dm^{-2}.

The incorporation of Al_2O_3 or TiO_2 particles strongly affects the microstructure of the metal matrix. As a result of the nanoparticle codeposition the microstructure of the nickel matrix changed drastically to the UD (unoriented dispersion) type [16]. An example is illustrated in Fig. 43 for a nanocomposite film prepared with the IJE system from a sulfamate electrolyte containing 60 g l^{-1} of 50 nm Al_2O_3 particles. High resolution SEM micrographs of that cross sectional sample were taken to examine the distribution of particles within the layers (Fig. 43b and c). The alumina particles appear as light spots in the darker nickel matrix, as validated by EDX spectroscopy. The particles appear to be incorporated in the nickel matrix preferentially as agglomerates (Fig. 43b and c) with a size of about 100 to 500 nm. A detailed characterization of the particle incorporation behavior and an examination of the particle-metal interface will be provided in the following section (section 3.6.3).

Figure 43   Cross-sectional SEM images of a Ni-Al$_2$O$_3$ composite film plated at 10 A dm^{-2}, 6.5 l min^{-1} and 60 g l^{-1} of 50 nm Al$_2$O$_3$ particles in the electrolyte. (with increasing resolution).

In the case of the pyrophosphate bath the pure nickel films as well as the nanocomposites exhibited a granular structure belonging to the UD (unoriented dispersion) type (Fig. 44). In accordance with the observation of the surface morphology, a general observation was that the codeposition of TiO$_2$ particles resulted in a refinement of the nickel grains.

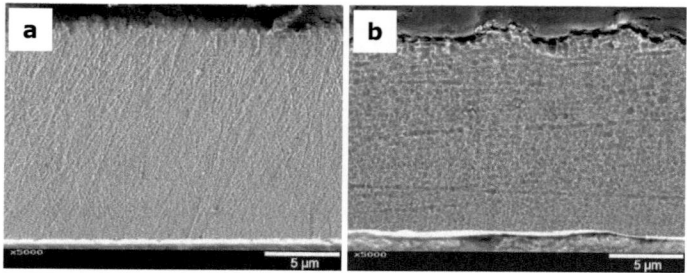

Figure 44   Cross sectional SEM image of pure nickel (a) and a Ni-TiO$_2$ nanocomposite (b) film deposited from pyrophosphate bath. (a) 10 A dm^{-2}; (b) 5 A dm^{-2}, 10 g l^{-1} TiO$_2$ in the electrolyte.

Of course, SEM does not provide the best microstructural analysis. For an extensive characterization of the grain size distribution and the crystallographic texture TEM and XRD investigations were performed. The results will be discussed in the following.

The XRD patterns of Ni films and Ni-TiO$_2$ composite films deposited at a current density of 1 A dm^{-2} are shown in Figs. 45a and b, respectively. The corresponding relative texture coefficients RTC$_{200}$ and RTC$_{111}$ were calculated according to equation 7 (p. 38) and are given in Table 5. The weak reflection at 2-Theta = 25.3° can be assigned to the (100) line of anatase (JCPDS no. 21-1272). The nickel matrix exhibited a face-centered cubic (fcc) lattice with different preferred orientations which were mainly influenced by the plating current density and the incorporation of alumina nanoparticles.

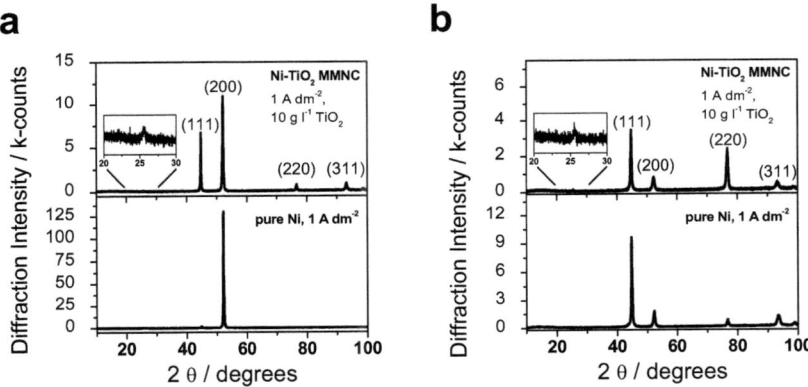

Figure 45 X-ray diffraction patterns of Ni and Ni-TiO$_2$ films plated at a current density of 1 A dm^{-2}. (a) acidic sulfamate bath, (b) alkaline pyrophosphate bath.

The XRD patterns of pure nickel deposits from the sulfamate bath revealed a strong (100) preferred crystalline orientation [133] detected by the intensive (200) reflection, as presented in Fig. 45a. A relatively strong <111> texture has been found to be associated with larger grains [133]. The increase of grain size with the current density has been also observed in the SEM micrographs of the surface morphology of pure nickel films [193]. However, the quality of <100> texture, *i.e.* the absolute value of the RTC$_{111}$ was significantly affected by the working parameters. An increase in the plating current density as well as the presence of the titania nanoparticles in the bath promoted a loss of texture, reflected by the reduction of the line (200) and the relative enhancement of the (111), (220) and (311) lines (Fig. 45a and Table 5). The relative texture coefficient RTC$_{200}$ decreased by up to 58 %, whereas RTC$_{111}$ increased by 3 to 20 % (Table 5). Using the alkaline bath containing pyrophosphate leads to a nickel crystal growth in direction of the (111) plane (Fig. 45b). Similar to the sulfamate bath, both increasing current density and the particle incorporation caused a loss of texture, indicated by decreasing RTC$_{111}$ and increasing RTC$_{200}$ values (Table 5). Thus, the (220) diffraction peak intensity increased considerably (Fig. 45b).

Table 5 Relative texture coefficient RTC_{hkl} of the pure nickel and Ni-TiO$_2$ composite films deposited from electrolytes containing 10 g l^{-1} TiO$_2$.

	i [A dm^{-2}]	pure nickel		Ni-TiO$_2$ composite	
		Sulfamate	Pyrophosphate	Sulfamate	Pyrophosphate
RTC_{111}	1	0.5	30.4	14.7	14.7
	10	3.7	28.9	24.0	21.2
RTC_{200}	1	98.2	14.7	58.0	8.4
	10	92.5	21.2	34.0	11.6

The crystallite size of the nickel films was estimated from the diffraction line broadening using the Debye-Scherrer method (section 2.3.3) [164]. The crystallite size of the nanocrystalline pure nickel films varied between 14 and 192 nm, depending on the electrolyte composition and the working conditions (Table 6). Increasing the current density resulted in an enhanced crystallite size. Such behavior has been found to be unique for nanocrystalline coatings deposited from an acidic nickel bath [133, 193, 194]. The width of the diffraction peaks of the Ni-TiO$_2$ nanocomposites is broader compared to thar of the pure nickel matrix (Fig. 45). This behavior is attributed to a decrease in the nickel crystallite size caused by the codeposition of the titania particles in the nickel matrix (Table 6). The structure and grain size of an electrodeposited layer is defined by the competition between nucleation and crystal growth [68]. It has been reported that the presence of nanoparticles provides more nucleation sites by increasing the surface area of the cathode and thus perturbs the nickel growth [195]. The resulting nanocomposite coatings show a smaller nickel crystallite size (Table 6) and a random orientation of the nickel growth (Fig. 45) [195].

Table 6 Crystallite size of pure nickel and Ni-TiO$_2$ composites, deposited from electrolytes containing 10 g l^{-1} nanoparticles, determined by the Scherrer line broadening using the (200) reflection of the XRD pattern.

	Sulfamate bath		Pyrophosphate bath	
	1 A dm^{-2}	10 A dm^{-2}	1 A dm^{-2}	10 A dm^{-2}
Pure nickel	170	192	18	14
Ni-TiO$_2$	42	60	16	12

Furthermore, XRD was used to characterize the preferred orientation and crystallite size of the Ni-Al$_2$O$_3$ composites plated with the IJE system. Similar to the Ni-TiO$_2$ films, the incorporation of alumina nanoparticles with a primary particle size of 50 nm leads to a distinct attenuation of the (200) reflection combined with a relative enhancement of the (111), (220) and (311) lines [181]. A comparable trend of the textural modification of the nickel matrix with a variation in the current density and as a result of the particle incorporation was found for the PP and PRP experiments, respectively.

Using the IJE system, the crystallite size of a pure nickel film deposited at a flow rate of 2.5 l min^{-1} and a current density of 10 A dm^{-2} was about 115 nm. In comparison a nickel alumina composite plated under the same working conditions from an electrolyte containing 120 g l^{-1} showed a nickel crystallite size of about 30 nm. These crystallite sizes are consistent with those determined for the pure nickel films plated with the PPE system as well as with those of the Ni-TiO$_2$ composite films (Table 6). It is interesting to note, that high amounts of particle incorporation are not directly correlated with a small nickel crystallite size. Rather, the crystallite size depends upon a variety of additional plating parameters particularly the plating current density (Table 6).

3.6.3 Particle Incorporation Behavior

Materials properties of MMNC films depend mainly on particle distribution, their bonding to the matrix and their influence on the microstructure of the metal matrix (section 1.5) which has been characterized by TEM (section 2.3.2), STEM, SEM, QBSD and EBSD (section 2.3.1.2). In this section, the electron microscopy results will be discussed for two selected composite samples which have been prepared from a sulfamate bath containing 10 g l^{-1} 13 nm Al$_2$O$_3$ particles using the PPE system and either DC or PRP (section 2.1.3.1). While the DC sample was prepared at 10 A dm^{-2}, the PRP sample was produced at i_p=5 A dm^{-2}, t_{on}=400 ms and i_{an}=1 A dm^{-2}, t_{rev}=20 ms. For comparison, a pure nickel film plated by PRP (same conditions as described above) was analyzed. The crystallite size of the nickel films was estimated from the diffraction line broadening using the Debye-Scherrer method (section 2.3.3) [164] as well as the electron backscattered diffraction (EBSD) analysis. Typically, EBSD is a technique used for microstructural characterization in materials science [160, 161]. However, compared to X-ray diffraction experiments (section 2.3.3) which supplies wide-ranging texture information of the whole film, EBSD illustrates local details of crystal growth at micron size [196]. Thus, this technique is advantageous in studying the metal growth at the nickel-substrate interface as well as around the co-deposited nanoparticles. Hence, the EBSD results have not been presented in the previous section (section 3.6.2) because of their relevance to the description of the particle incorporation behavior in the metal matrix which will be discussed in the present section. EBSD maps were calculated after a minor clean up excluding grain dilatation but comprising grain confidence index (CI) standardization with 2° tolerance angle and neighbour CI correlation and show the correctly indexed fraction with CI > 0.1. Mean grain size parameters were derived from EBSD quality maps from the lower part of the layer near the substrate and at the film surface. Twin boundaries were considered and excluded for grain size evaluation.

Table 7 summarizes the grain features derived from XRD and EBSD studies. The grain sizes determined from the EBSD measurements are obviously larger compared to those computed from the XRD patterns. The variations might be caused by the fact that rather often polycrystalline samples exhibit variations in the crystallite size with grains including dislocations, stacking faults, twin planes and further lattice defects which complicate the interpretation of the XRD peak broadening [159]. The crystallite size determined by XRD analysis corresponds to the size of the coherently diffracting domains [159]. In general, this value is obviously smaller compared to the grain size obtained by electron microscopy [159]. Furthermore, the values derived from XRD measurements comprise even grains below 100 nm in contrast to the EBSD results due to the scanning step size of 100 nm. Hence, it has to be carefully specified which experimental technique has been used for determination of the crystrallite/grain size in order to enable a comparison. Nevertheless, both values show the same tendency.

Table 7 Grain features of the nickel matrix.

Sample	Mean grain size [nm]			Particle incorporation [vol-% Al_2O_3]
	XRD	EBSD (initial and surface region)	EBSD (only surface region)	
Pure nickel (PRP)	192	360±390	430±450	0
Pure nickel (DC)	317	Not measured	Not measured	0
Composite (PRP)	231	410±560	420±420	10.3
Composite (DC)	184	310±290	390±370	2.5

Back scattered electron detection (QBSD) revealed as an appropriate tool for the visualization of the nickel crystal growth by means of orientation contrast as well as for particle distribution by means of material contrast. The layer growth direction is upwards in all cases. The QBSD image taken from the sample prepared by PRP (Fig. 46a) confirms the field-oriented growth of nickel films from the sulfamate bath as detected by XRD (Fig. 45a, p. 69) and cross sectional SEM micrographs (Fig. 42, p. 67). A distinct <100> fibre texture in growth direction is illustrated in the nickel crystal direction map color coded by the inverse pole figure (IPF) (Fig. 46b). Starting with a nearly 1 μm thick fine crystalline initial layer, the subsequent layer is formed by columnar grains with growing size from ~0.4 μm inside the film to ~1.1 μm at the film surface. The formation of an initial layer with tiny grains in the vicinity of the substrate has been previously reported for nickel and nickel composites plated from an acidic Watts electrolyte [196]. In these investigations, particle reinforcement included Al_2O_3, TiO_2 or SiC with different sizes ranging from 21 nm to 20 μm.

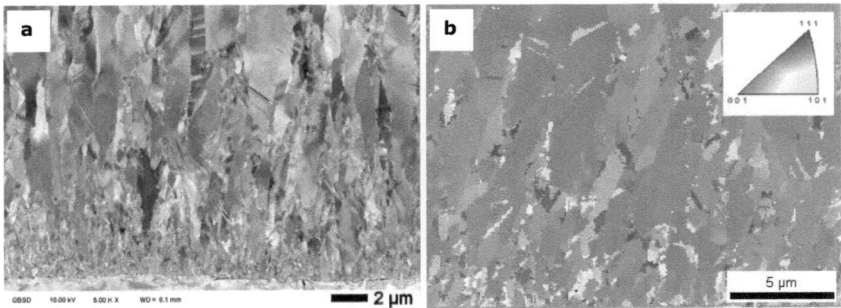

Figure 46 QBSD image (a) and IPF color coded nickel map (b) of a pure nickel film deposited from the sulfamate electrolyte by PRP: $i_P=5$ A dm^{-2}, $t_{on}=400$ ms and $i_{an}=1$ A dm^{-2}, $t_{rev}=20$ ms.

The effect of alumina nanoparticle incorporation on the texture and grain structure of the nickel matrix as well as a rough indication of the particle distribution in the metal matrix can be derived from the QBSD maps shown in the Figs. 47a and 48a.

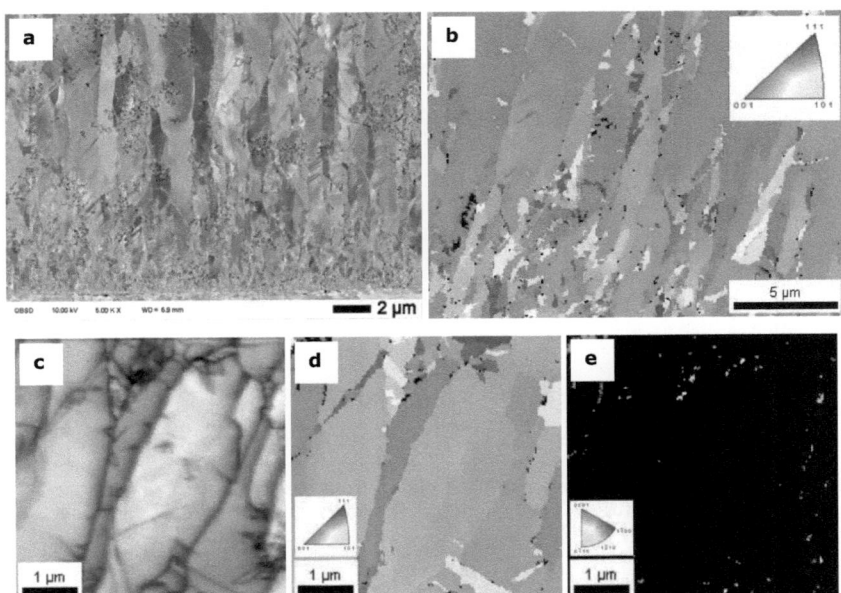

Figure 47 QBSD image (a) and IPF color coded nickel map (b) as well as detail in image quality map (c) IPF color coded nickel map (d) and IPF color coded alumina map (e) of a Ni-Al$_2$O$_3$ film deposited by PRP (same plating conditions as Fig. 46, 10 g l^{-1} 13 nm Al$_2$O$_3$ in the electrolyte).

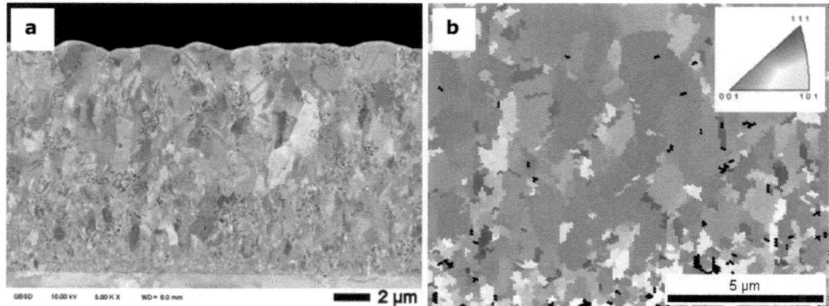

Figure 48 QBSD image (a) and IPF color coded nickel map (b) of a Ni-Al$_2$O$_3$ film deposited from the sulfamate electrolyte containing 10 g l^{-1} 13 nm Al$_2$O$_3$ by DC: i=10 A dm^{-2}.

Moreover Fig. 49 gives the EBSD pole plots measured in the cross section of two Ni-Al$_2$O$_3$ composites prepared under DC and PRP conditions. These plots show a distinct <100> fibre texture in film growth direction with the (100) spots scattered in the centre and in a broadened circle. Comparison of the two pole figures, leads to the assumption that PRP conditions (Fig. 49b) promote the columnar growth, the twin formation and a sharper texture.

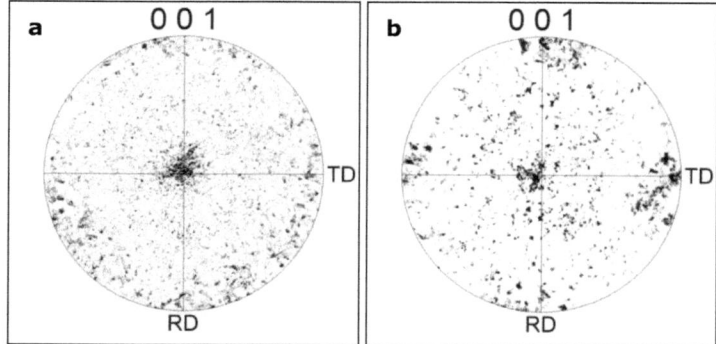

Figure 49 Pole figures of composite films plated from electrolyte containing 10 g l^{-1} 13 nm Al$_2$O$_3$. (a) DC: 10 A dm^{-2}; (b) PRP: i$_P$=5 A dm^{-2}, t$_{on}$=400 ms and i$_{an}$=1 A dm^{-2}, t$_{rev}$=20 ms.

The alumina particles which are mainly present in the form of agglomerates appear as dark spots in the lighter nickel matrix. Using EDX analysis in the cross section (section 2.3.1.2), the volume fraction of alumina particles in these films was determined to be ~10.3 vol-% (PRP sample; Fig. 47), and ~2.5 vol-% (DC sample; Fig. 48). In good agreement with the pure nickel film (Fig. 46a) and the findings in Ref. [196] the composite film formation starts with an initial layer of tiny grains adjacent to the substrate. With increasing distance from the Cu substrate, columnar nickel grains appeared. However, the columnar nickel grains seem to be refined due to the presence of the particles (Figs. 47a and 48a). Regardless of the respective current modulation used, the black spots in the IPF color coded nickel maps of the composite films (Figs. 47b and 48b) represent highly disordered parts of the nickel matrix probably containing alumina particles. Although there is a distinct ambiguity in this interpretation, further prove of localization of alumina particles comes from the image quality, nickel (Figs. 47b,d and 48b) and alumina orientation maps (Figs. 47e) shown, which were taken with a reduced step size of 50 nm.

The image quality map (Fig. 47c) represents low ordered regions in the sample, *i.e.* nickel grain boundaries and dark spots in the vicinity of grain boundaries as well as in the middle of the central grain. A few of them are mirrored in the orientation images of nickel (Fig. 47d) and of alumina (Fig. 47e). Nevertheless, no alumina was detected in the central pink grain, which is due to the particle size and the EBSD step size. Even so, both the QBSD and the EBSD maps indicate that the alumina nanoparticles are predominantly incorporated in the grain boundary zone. It is apparent that the alumina particles terminate the columnar nickel crystals and consequently affect the growth direction which is consistent with the XRD results (Fig. 45a and Table 5, p. 69).

The detailed incorporation behavior of alumina particles, *i.e.* the size and distribution of co-deposited particles in the nickel matrix and the particle-nickel bonding, was studied using QBSD, EsB, STEM and TEM imaging. The results will be discussed in the following for the composite film plated by PRP. In QBSD imaging the alumina particles appear as dark spots. The alumina particles agglomerate and form chains of particles as detected by QBSD imaging (Fig. 50a). The resolution in back scattered electron imaging is limited by their relatively high escape depth. Resolution can be enhanced using only elastically scattered back scattered electrons (EsB) with a high angle to the surface (Fig. 50b) which in addition reveals fine twin lamellae in the nickel grain, as well as imaging of electron transparent samples with STEM. In dark field imaging the particles appear as black and dark gray spots in the vicinity of small grains (Fig. 50c). However, the grain refinement by particle incorporation is lower compared to the grain refinement in the interface region (Fig. 50d). Nevertheless, the resolution in SE imaging is insufficient to study the particle-metal interface.

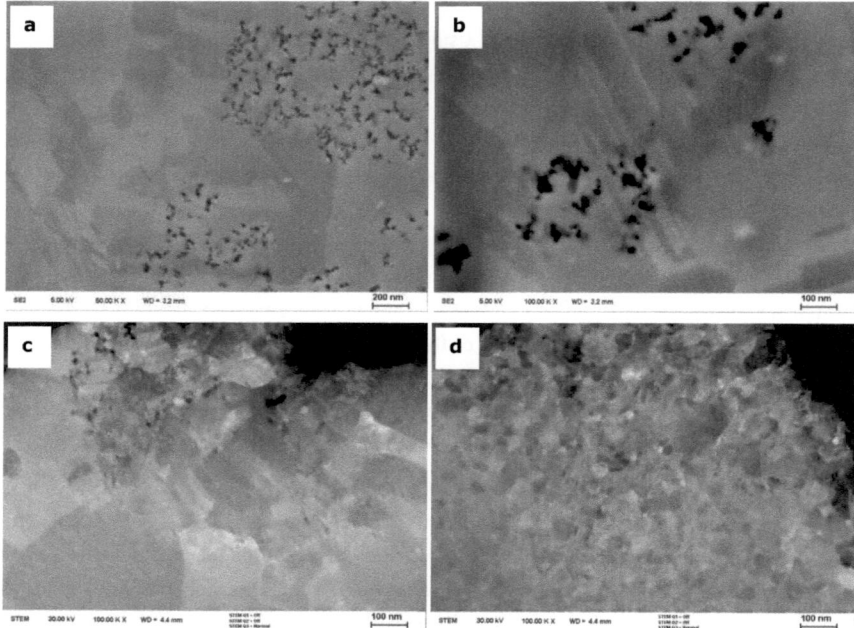

Figure 50 SE (a, b) and STEM (c, d) images of a Ni-Al$_2$O$_3$ film deposited by PRP (same plating conditions as Fig. 46, 10 g l^{-1} 13 nm Al$_2$O$_3$ in the electrolyte).

Transmission electron microscopy at 200 kV shows numerous chains of agglomerated alumina particles and sometimes single particles in the polycrystalline nickel matrix as bright spots in bright field imaging (Fig. 51). These chain-like agglomeration already originates from the cooling process during particle preparation (section 2.1.2.1) as can be seen in Fig. 15 (p. 41) and was confirmed by the producer. Compared to SE imaging, the alumina particles can hardly be distinguished from the matrix in TEM due to the better transparency of the matrix at higher acceleration voltage. Hence, some of the incorporated particles are marked in Fig. 51.

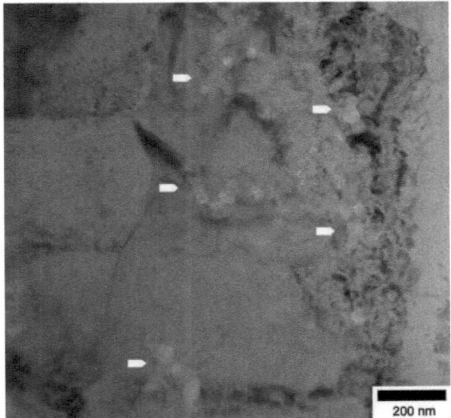

Figure 51 TEM image of a Ni-Al$_2$O$_3$ film produced by PRP (same plating conditions as Fig. 46, 10 g l^{-1} 13 nm Al$_2$O$_3$ in the electrolyte).

Nevertheless, particle incorporation can also be confirmed by defocusing the TEM showing a lesser delocalisation compared to nickel crystals. By further increasing the magnification in TEM imaging the incorporated individual particles are characterized by a nearly spherical shape and a size of ~10-20 nm.

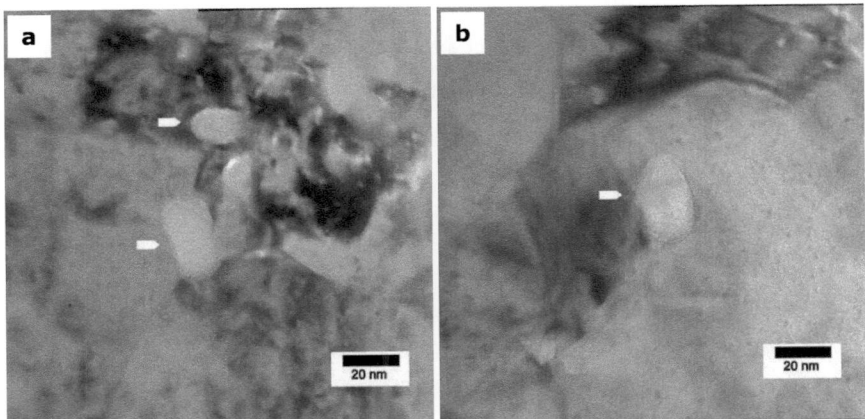

Figure 52 High resolution TEM images of a Ni-Al$_2$O$_3$ film produced by PRP (same plating conditions as Fig. 46, 10 g l^{-1} 13 nm Al$_2$O$_3$ in the electrolyte).

Also, a good particle matrix bonding without any voids can be confirmed by TEM at higher magnification (Figs. 52 and 53). Nevertheless, in high resolution (Fig. 53) no lattice order can be observed for the particles despite the clear image of {111} lattice planes with 0.203 nm spacing and {110} lattice planes with 0.25 nm spacing of the nickel matrix. Moreover, the lack of lattice planes of the alumina particles is probably due to the preparation state of the sample which will be improved in future. Finally, both EBSD and TEM studies reveal that both agglomerated as well as individual alumina particles are preferentially incorporated in the vicinity of grain boundaries.

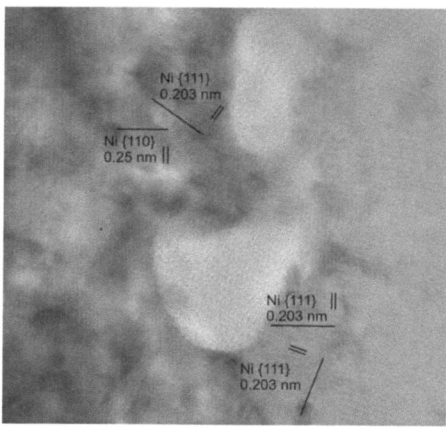

Figure 53 High resolution TEM image of a Ni-Al_2O_3 film (same plating conditions as Fig. 46, 10 g l^{-1} 13 nm Al_2O_3 in the electrolyte) including the subscription of certain lattice planes of the nickel matrix.

3.6.4 Mechanical Properties

3.6.4.1 Vickers Microhardness

Vickers microhardness measurements were performed for pure nickel and particle reinforced films (section 2.3.4). The microhardness of the coatings varied from 210 HV for pure nickel to 580 HV for composite films. The literature values of pure Ni films from the acidic sulfamate electrolyte are 240-370 HV [132] and from the alkaline pyrophosphate bath 380-450 HV [197], depending on the working parameters. The hardness of metal films and composite coatings is related to the microstructure as well as to the quantity and distribution of the reinforcement phase (section 1.5).

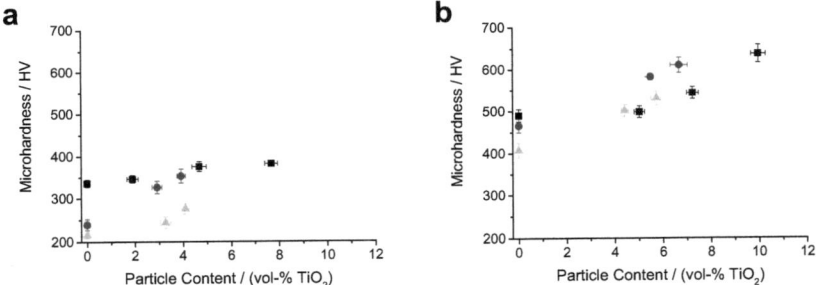

Figure 54 Correlation between the Vickers microhardness and the TiO_2 content of the layer; (■) 1 A dm^{-2}; (●) 5 A dm^{-2}; (▲) 10 A dm^{-2}. (a) sulfamate electrolyte; (b) pyrophosphate electrolyte.

Figure 54 indicates that the hardness of the pure nickel films showed a tendency to increase with decreasing current density. Due to a variation of the current density the nickel growth mode and crystallite size was significantly altered (Fig. 45, p. 69). For the alkaline pyrophosphate bath, an increase from 1 to 10 A dm^{-2} led to a decrease of the crystallite size by about 40% depending on the plating bath and the operating conditions (Table 6, p. 70), as determined by the Scherrer line broadening of the nickel diffraction lines (Fig. 45, p. 69). Hence, the hardness increase of the nickel and nickel composite films is mainly related to the Hall-Petch hardening effect of the ultrafine grains of the nanocrystalline nickel matrix [198, 199]. Regardless of the composition of the nickel plating bath and the applied current density, the microhardness of the nickel coatings increased significantly due to the incorporation of titania nanoparticles (Fig. 54). On the one hand the increase of the hardness can be explained by a decrease of the nickel crystallite size due to the presence of the TiO_2 nanoparticles in the bath (Table 6, p. 70).

Furthermore, the enhanced hardness might be caused by the high hardness and strength of the titania particles. The incorporation of nanoparticles in a nickel matrix is known to inhibit the plastic flow of the metal (section 1.5) [200].

The microhardness of the nickel coatings plated under PP and PRP conditions varied between 250 and 520 HV, whereas the hardness values of pure nickel films were between 250 and 380 HV [117]. Thus, in accordance with Refs. [201, 202] it appears that both PP and PRP conditions (compared to DC conditions) have only little impact on the hardness of nickel coatings from the sulfamate bath. In agreement with Ref. [75] the microhardness of pure nickel films and nanocomposites was found to increase with decreasing duty cycle [117]. These results are in good agreement with the effects of PP conditions on the surface morphology of the coatings (Fig. 41, p. 66). It is well known that the hardness of nickel films plated from the acidic sulfamate bath under DC conditions increases with decreasing plating current density [83]. In PP experiments the average plating current density (Eq. 8, p. 55) increases with decreasing duty cycle and increasing pulse frequency. Hence, the improved hardness of the nickel films due to a variation of the PP plating parameters can be explained by the change of the average current density, which is in good agreement with the results of DC plating [83]. Additionally, a lower duty cycle corresponds to a relatively longer pulse pause that leads to an enhanced incorporation of alumina particles within the metal matrix (Fig. 30, p. 54).

Finally, the hardness of the nickel nanocomposites is only slightly affected by the absolute amount of incorporated alumina particles. The main parameter governing the hardness of the coatings is the structure of the metal matrix that is mainly influenced by the plating conditions, viz. average plating current density. As a result of the particle incorporation, the microstructure of metal films plated from acidic electrolytes changed from the FT-type (Fig. 42, p. 67) to the UD-type (Figs. 43 and 44, p. 68), which is known to improve the hardness of the coating considerably [16].

3.6.4.2 Abrasion Resistance

Wear resistance is an important factor governing the industrial application of surface coatings [203]. The wear performance of electroplated composite films depends on the microstructure of the metal matrix as well as on the amount of incorporated particles [204].

Results and Discussion

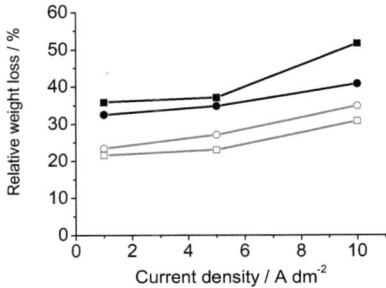

Figure 55 Relative weight loss after bi-directional sliding test as a function of the plating current density. Sulfamate electrolyte: (■) 0 g l^{-1}; (●) 10 g l^{-1} TiO$_2$; Pyrophosphate electrolyte (□) 0 g l^{-1}; (○) 10 g l^{-1} TiO$_2$.

The relative weight loss after bi-directional sliding test of Ni and Ni-TiO$_2$ nanocomposite films is shown in Fig. 55. As is evident from this figure, the wear of the films increased with increasing current density. This behavior correlates with the decrease of the microhardness with increasing current density (Fig. 54, p. 79). Hence, it can be explained by the strong preferred orientation (Fig. 45 and Table 5, p. 69) and the small crystallite size of the nickel matrix (Table 6, p. 70) [204]. The best abrasion resistance was observed for the coatings deposited from the alkaline pyrophosphate bath (Fig. 55), which also reveal the higher microhardness (Fig. 54, p. 79) [204].

3.6.5 Magnetic Properties

The magnetic properties of pure Ni and Ni-Co/Ni-Fe$_3$O$_4$ coatings were analyzed by recording hysteresis loops (section 2.3.6). In general the magnetization curve of a material depends on the composition and the shape of the sample [166].

The saturation moment M_s of the pure Ni coatings was found to decrease with increasing plating current density (Table 8). The elemental analysis using EDX showed only Ni and Cu (substrate) within the pure Ni films. The differences in the saturation moments can be explained by the differences in the deposited Ni mass. This is in good agreement with the film thickness measured by means of XRF spectroscopy. The saturation magnetization of the pure Ni coatings calculated from the film thickness varied between 35 and 77 mT. The generally accepted value for pure Ni is 55.2 mT [132]. Variations are caused by the determination of the deposited Ni mass because the surface morphology and porosity of the coatings has not been taken into account.

Table 8 Magnetic properties of pure Ni and Ni-Co composites. Composites have been prepared from electrolytes containing 6 g l^{-1} Co particles. Particles: c-cubic, d-discoid.

i [A dm^{-2}]	B	particles	H$_{mes}$	H$_c$ [A m^{-1}]	M$_r$/M$_s$	M$_s$ [A m^2]	vol-% Co
0.5	0		x, y	4.85	0.12	16.7	
1	0		x, y	5.81	0.12	10.7	
2	0	no	x, y	5.17	0.15	8.0	/
5	0		x, y	5.65	0.21	2.5	
10	0		x, y	5.81	0.34	1.1	
1	0		x, y	3.02	0.20	33.6	34
5	0		x, y	4.22	0.27	10	30
10	0		x, y	4.77	0.42	4.5	26
1	z	6 g l^{-1} c	x, y	2.07	0.85	2.6	35
5	z		x, y	1.83	0.75	2.3	35
10	z		x, y	1.75	0.65	2.8	29
5	x		x	3.34	0.80	1.5	24
5	x		y	2.79	0.65	1.5	24
5	z		x, y	2.07	0.76	3.3	41
10	z	6 g l^{-1} d	x, y	2.63	0.82	1.8	36
5	x		x	2.55	0.79	2.6	25
5	x		y	2.07	0.45	2.6	25

The remanence of the pure nickel coatings is relatively low (Table 8). The presence of a 100 mT magnetic field during the electrodeposition process has almost no effect on the hysteresis loops (Fig. 56).

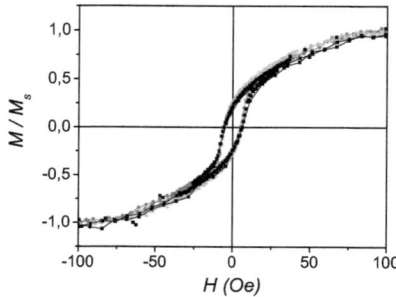

Figure 56 Hysteresis loops of pure Ni films deposited at 5 A dm^{-2}. (■) **B** = 0 mT; (●) **B** = 100 mT (z-direction); (▲) **B** = 100 mT, (x-direction, Fig. 9, p. 28).

In marked contrast to the pure Ni films, the remanence of the Ni-Co composite films is significantly affected by the presence of a static magnetic field during the ECD process (Table 8). The Co nanoparticles are expected to show a tendency to align parallel with the magnetic field direction during ECD. Finally, the particles would become fixed in the deposited nickel matrix. If the orientation of the field is in z-direction, the particles are most probably orientated with their magnetization in that direction. Hence, there is no preferential direction for the measurement of hysteresis loops in the x-y-plane. In the case of a magnetic field oriented in the x-direction during ECD, the Co nanoparticles are expected to orientate with their magnetization also in x-direction. Thus, there should be a preferential direction of the hard magnetic particles in the electrodeposited composite film, which can be proved by differences in the magnetic properties of the coating depending on the direction of the external magnetic field during VSM measurements. Hysteresis loops recorded in x-direction show a higher magnetic hardness than in y-direction (Table 8). The effects found in the magnetization properties of the Ni-Co composites are almost independent from the shape of the co-deposited Co nanoparticles; there are small quantitative differences between discoidal and cubic Co particles (Table 8). Furthermore, the orientation and distribution of the Co particles in the Ni matrix has to be considered. It cannot be expected that the Co nanoparticles are randomly distributed in the nickel matrix. They might have a tendency to align and stick together.

Likewise to the pure Ni films (Table 8), the Ni-Fe$_3$O$_4$ composites showed a gradual decrease of the saturation moment with increasing plating current density (data not shown). The highest values of the saturation moment were found for the Ni-Fe$_3$O$_4$ composites plated in the presence of a magnetic field parallel to the electrical field. The coercive field for all layers is relatively low, between 1.1 and 5.8 A m^{-1}. For the Ni-Co composites deposited under the influence of a static magnetic field the coercive field is lower than for the same deposition without magnetic field (Table 8). An almost opposite trend was found for the Ni-Fe$_3$O$_4$ composites, where the coercivity was higher due to the presence of the magnetic field during electrodeposition (data not shown).

Recently, magnetron sputtered rare earth (RE)-Fe-B thin films have been utilized for magnetic micro-electromechanical system (MEMS) applications [36]. Compared to the composite layers discussed here the magnetic properties of RE-Fe-B films are superior with regard to MEMS application [205-208]. However, the preparation of hard magnetic thin films by means of ECD has several advantages compared to other methods, e.g. uniformity and simplicity even for complex shapes, low levels of contamination, easy control of the process and the ability of continuous processing [85]. Future work should be directed towards the improvement of the magnetic properties of the co-deposited films. In fact, a variation in the particle size and shape as well as in the composition of the nanocomposite films might have a deep impact on the magnetic properties.

4 Model of Electrocodeposition using an Impinging Jet Electrode

Based on the experimental results from ECD of 50 nm alumina particles in nickel with an IJE system (section 3.4), a kinetic model is developed to predict the volume fraction of particles in the composite film. In this model, the basic approach of Vereecken et al. (section 1.3.4) [51, 114] was modified by replacing the transport parameters of the RDE by those of the unsubmerged IJE system (section 2.1.3.2).

4.1 Mathematical Model

In this model, the motion of particles within the hydrodynamic boundary layer, δ_0, of an IJE system is described via an analysis of the total force acting on a single particle (Fig. 57). In front of the impinging zone of the IJE system, one can distinguish four main regions [115, 209]: the diffusion boundary layer for particles ($0<z<\delta_p$), the diffusion boundary layer for reactive ions ($0<z<\delta$), the viscous sublayer ($0<z<\delta_0$) and the free jet flow region ($z>\delta_0$), where both shear stress and viscous forces are negligible [209, 210]. The concentration of the reactive ionic species at the electrode surface and in the bulk solution are $c_{i,s}$ and $c_{i,b}$, and $c_{p,s}$ and $c_{p,b}$ are the corresponding bulk concentration and surface concentration of the particles. A variation of the particle concentration is assumed to be confined within the particle diffusion boundary layer, δ_p.

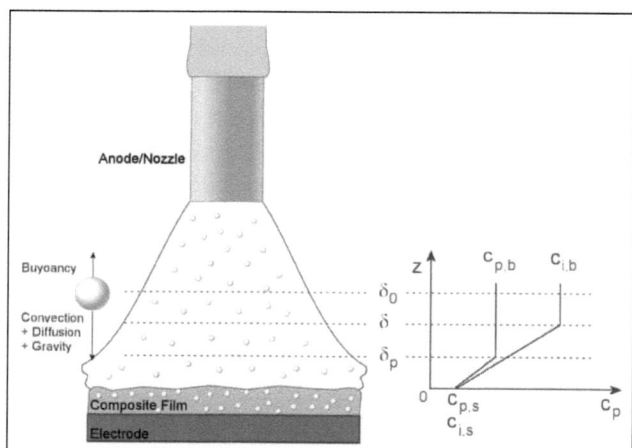

Figure 57 Schematic representation of the particle concentration in the impinging region of the unsubmerged IJE system. $c_{p,b}$, $c_{i,b}$ are the bulk concentration and $c_{p,s}$, $c_{i,s}$ are the concentration at the electrode surface of the particle and reactive metal ion, respectively.

Following the development of the model of Vereecken et al. [51, 114], the total force acting on a single rigid particle in an unbounded homogenous solvent is given by the product of the net drift velocity of the particles (v_{Dr}) times the friction coefficient (f) [114].

$$F_{total} = f \cdot v_{Dr} \tag{9}$$

According to Stokes' law the friction coefficient can be calculated by $f = 6\pi\eta r$, where η is the viscosity of the solution and r the particle radius.

The total force on the particle (F_{total}) in an impinging jet is the sum of convectional diffusion (F_{Diff}) and gravitation (F_{Grav}) (Fig. 57). Of course the gravitational term has to be corrected for the influence of the buoyancy of the dispersed nanoparticles. Hence, the total force exerted on a particle in front of the electrode can be written as

$$-F_{total} = F_{Diff} + F_{Grav} . \tag{10}$$

The diffusion force is known to be the negative of the chemical potential gradient ($\nabla\mu$), which corresponds in an ideal solution to [101]

$$F_{Diff} = -\nabla\mu = -\frac{k_B T}{c_{p,b}} \nabla c_p \tag{11}$$

where k_B is Boltzmann's constant and T is the temperature. The orientation of the driving force of the diffusion process, the concentration gradient (∇c_p), is schematically shown in Fig. 57. It is characterized by a linear concentration drop within the diffusion layer thickness, δ_p.

The second component is due to gravity and is given as [101]

$$F_{Grav} = -mg = -\frac{4\pi r^3}{3}(\rho_p - \rho_f)g \tag{12}$$

where m is the mass of one particle, g is the acceleration due to gravity, r is the particle radius, ρ_p and ρ_f are the density of the particle and the fluid, respectively. From Eq. 12 it can be seen that the gravitational force is a function of the particle radius.

Substituting Eq. 9, 11 and 12 in Eq. 10, the total force acting on a particle in front of the electrode can be written as

$$F_{total} = \frac{k_B T}{c} \nabla c_p + \frac{4\pi r^3}{3}(\rho_p - \rho_f)g = 6\pi\eta r v_{Dr} \tag{13}$$

The particle flux is defined as $J_p = c_p \cdot v_{Dr}$. Hence, substituting for the velocity Eq. 13 can be rewritten as

$$J_p = \frac{k_B T}{6 \pi \eta r} \nabla c_p + \frac{2(\rho_p - \rho_f) g r^2}{9 \eta} c_p \qquad (14)$$

Comparison to Fick's first law shows that $k_B T / 6 \pi \eta r$ is the particle diffusion coefficient D_p [101].

Following the model of Vereecken et al. [51, 114] we will use the Nernst diffusion layer model to describe the particle flux to the electrode [102]. The driving force for the diffusion process is characterized by a linear concentration drop within the diffusion layer thickness, δ_p, and is given by

$$\nabla c_p = \frac{(c_{p,b} - c_{p,s})}{\delta_p} \qquad (15)$$

where $c_{p,b}$ is the bulk particle concentration, $c_{p,s}$ particle concentration at the electrode surface, and δ_p the thickness of the diffusion boundary layer for particles.

According to the diffusion theory, the average particle concentration within the diffusion layer is given by $c_p = (c_{p,b} + c_{p,s})/2$. Thus, the particle flux (Eq. 14) can be written as

$$J_p = D_p \frac{(c_{p,b} - c_{p,s})}{\delta_p} + \frac{(\rho_p - \rho_f) g r^2}{9 \eta} (c_{p,b} + c_{p,s}) \qquad (16)$$

The volume fraction of particles (x_V) within the electroplated composite film is given by

$$x_V = \frac{n_p N_A V_p}{n_p N_A V_p + n_m V_{m,M}} \qquad (17)$$

where n_p and n_m are the number of moles of particles and metal atoms in the film, V_p is the particle volume, $V_{m,M}$ is the molar volume of the metal matrix, and N_A is Avogadro's number. The ratio of the number of moles of particles to those of metal atoms n_p / n_m is equal to the ratio of their fluxes J_p / J_m. During the process of electrocodeposition, both J_p and J_m are supposed to be positive. Noting that the flux of metal ions to the electrode surface is given by $J_m = i / z F$, where i is the applied current density, z is the charge number of the species and F is Faraday's constant, Eq. 17 can be written as

$$J_p = \frac{3 V_{m,M}}{4 \pi r^3 z F N_A} \frac{x_V}{1 - x_V} i \qquad (18)$$

Combining Eq. 16 and 18 under consideration of $V_p / V_m = x_V / (1 - x_V)$ gives an expression for V_p / V_m as a function of the plating parameters

$$\frac{V_p}{V_m} = \frac{4\pi r^3 zFN_A}{3V_{m,M} i} \cdot \left(D_p \frac{(c_{p,b} - c_{p,s})}{\delta_p} + \frac{(\rho_p - \rho_f)gr^2}{9\eta}(c_{p,b} + c_{p,s}) \right) \quad (19)$$

In the case of a sufficient high metal deposition rate, all particles arriving at the electrode surface are incorporated in the metal matrix. If the probability of particle entrapment is unity and the particle incorporation is limited by the particle diffusion, the particle concentration at the electrode surface is $c_{p,s} = 0$ and Eq. 19 can be written as

$$\frac{V_p}{V_{metal}} = \frac{4\pi r^3 zFN_A}{3V_{m,M} i} \cdot \left(\frac{D_p}{\delta_p} + \frac{(\rho_p - \rho_f)gr^2}{9\eta} \right) c_{p,b} \quad (20)$$

Chin and Hsueh extensively investigated mass transfer with an unsubmerged IJE system [121]. Using the Chilton-Colburn analogy [121], which utilizes a similarity between mass and momentum transfer to a flat plate, they obtained a correlation for the rate of mass transfer to the impinging region of the IJE system in the form

$$Sh = 0.9\, Re^{1/2}\, Sc^{1/3} \left(\frac{H}{d} \right)^{-0.09} \quad (21)$$

where the Sherwood number $Sh = kd/D_m$ is a function of the Reynolds number $Re = \upsilon d/\nu$, which is based on the diameter of the jet nozzle (d) and the exit velocity from the nozzle (υ), and the Schmidt number $Sc = \nu/D_m$, which is a ratio of the kinematic viscosity (ν) of the electrolyte to the diffusion coefficient of the reactive species (D_m), and the ratio of the distance between the nozzle and substrate (H) to the diameter of the jet nozzle (d). By defining the average mass transfer coefficient [102] as $k = D_m/\delta$ and using the Eq. 21, the diffusion layer thickness at an unsubmerged IJE can be calculated as

$$\delta = \frac{d}{0.9\, Re^{1/2}\, Sc^{1/3} \left(\frac{H}{d} \right)^{-0.09}} = \frac{\nu^{1/6} D_m^{1/3} H^{0.09} d^{0.41}}{0.9\, \upsilon^{1/2}} \quad [22]$$

Due to the small size of the alumina nanoparticles (50 nm), transport by Brownian diffusion is important. The diffusion coefficient of the alumina nanoparticles in the nickel sulfamate bath is calculated from the Stokes-Einstein equation [101] to be $D_p = 4.19 \cdot 10^{-12}\ m^2s^{-1}$ using the values given in Table 9. The diffusivity of the reactive species, Ni^{2+} ions, is known to be $D_m = 0.661 \cdot 10^{-9}\ m^2s^{-1}$ [211].

Table 9 List of values for physical properties.

Physical property	Value at 298 K	Reference
r	25 nm	[145]
$c(Ni^{2+})$ of bulk solution	1.08 M	[178]
$D_m = D(Ni^{2+})$	$0.661 \cdot 10^{-9}$ m² s⁻¹	[211]
$V_{m,M} = V_{m,Ni}$	$6.594 \cdot 10^{-6}$ m³ mol⁻¹	[211]
H (nozzle-substrate distance)	$6.5 \cdot 10^{-3}$ m	[178]
d (nozzle diameter)	$1.09 \cdot 10^{-2}$ m	[178]
η	2.09 g m⁻¹ s⁻¹	experimental
v	$1.61 \cdot 10^{-6}$ m² s⁻¹	experimental
ρ_p (= $\rho_{Alumina}$)	3842 kg m⁻³	[145]
ρ_f	1300 kg m⁻³	experimental
ρ_{Ni}	8900 kg m⁻³	[211]

Hence, the diffusion boundary layer thickness for the nanoparticles was calculated from Eq. 22, with D_p being substituted for D_m, such that

$$\delta_p = \frac{v^{1/6} D_p^{1/3} H^{0.09} d^{0.41}}{0.9 \upsilon^{1/2}} = 0.185 \cdot \delta \qquad (23)$$

The dependence of the particle diffusion boundary layer thickness on the electrolyte velocity is shown in Fig. 58 for the investigated flow rate range of the IJE system. As expected, an increase in the jet flow rate decreases the diffusion layer thickness. Of course, the diffusion boundary layer thickness of the Ni^{2+} ions also decreases with increasing electrolyte velocity (Eq. 23).

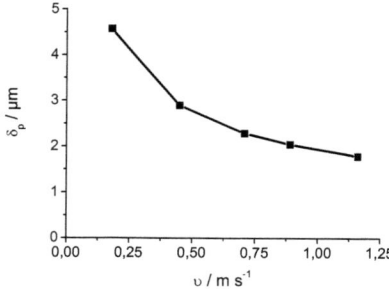

Figure 58 Particle diffusion boundary layer thickness as a function of the electrolyte velocity.

Substituting Eq. 23 into Eq. 20

$$\frac{V_p}{V_m} = \frac{4\pi r^3 z F N_A}{3 V_{m,M} i} \cdot (A_D + A_G) c_{p,b} \qquad (24)$$

where A_D and A_G are defined as

$$A_D = \frac{0.9 \upsilon^{1/2} D_p^{2/3}}{\nu^{1/6} H^{0.09} d^{0.41}} \qquad (25)$$

$$A_G = \frac{(\rho_p - \rho_f) g r^2}{9 \eta} \qquad (26)$$

The parameter A_D is related to the diffusion flux whereas A_G is related to the gravitational force corrected for the buoyancy. For a given ECD system, i.e. a given particle size and electrolyte composition, A_G is a constant and A_D is a function of the experimental parameters, particularly the jet flow rate.

The impact of the several forces acting on a particle (Eq. 10, p. 85) can be illustrated by plotting the magnitude of A_D and A_G as a function of the particle diameter (Fig. 59). The parameters were calculated using the IJE process parameters (section 2.1.3.2) and the values of the physical properties given in Table 9. As described above, the parameter A_G is a constant for a given particle-electrolyte combination. On the other hand A_D mainly depends on the IJE flow rate. However, the changes caused by a variation of the flow rate between 1 and 6.5 l min^{-1} are almost negligible with regard to the order of magnitude of the absolute value of the parameter A_D.

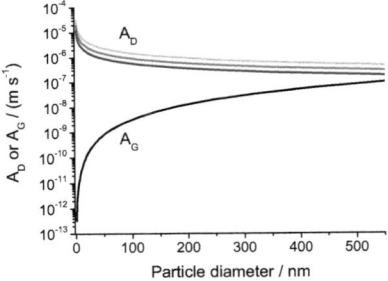

Figure 59   Parameter A_D (IJE flow rate: — 1 l min^{-1}, — 2.5 l min^{-1}, — 6.5 l min^{-1}) and A_G (—) in Eq. 24 vs. particle diameter.

As can be seen from Fig. 59, the impact of the gravitational force increases with increasing particle size. For particles with a diameter of about 50 nm (as used here) $A_G / A_D \leq 10^{-4}$, thus the influence of the gravitational force can be neglected. However, in the case of 500 nm particles, the values of the two parameters are almost in the same order of magnitude.

Finally, due to the predominance of the parameter A_D over the parameter A_G, it can be summarized that the convectional diffusion dominates over the gravitational effect in the case of 50 nm alumina particles (as used here). Therefore, Eq. 24 simplifies to

$$\frac{V_p}{V_m} = \frac{4\pi r^3 z F N_A}{3 V_{m,M} i} \cdot \frac{0.9 \, \upsilon^{1/2} D_p^{2/3}}{\upsilon^{1/6} H^{0.09} d^{0.41}} \cdot c_{p,b} \tag{27}$$

This equation can be rewritten in terms of the number of particles per unit volume of plated metal matrix, m_p, and the number density of particles (per unit volume of solution) in the electrolyte, n_b. Taking into account that $n_b = N_A \cdot c_{p,b}$, the following expression describes the number of co-deposited particles under diffusion limiting conditions.

$$m_p = \frac{z F}{V_{m,M} i} \cdot \frac{0.9 \, \upsilon^{1/2} D_p^{2/3}}{\upsilon^{1/6} H^{0.09} d^{0.41}} \cdot n_b \tag{28}$$

According to the Stokes-Einstein equation [101] the particle diffusion coefficient, D_p, depends mainly on the particle size. As a result, the absolute value of the parameter A_D, accounting for the particle transport by means of diffusion (Eq. 25), varies considerably as a function of the particle size. The nanoparticle diameter in the electrolyte can be different from that of the dry nanopowder. In general, the alumina nanoparticles are expected to aggregate in aqueous solutions, particularly in the plating bath with high ionic strength [101]. Hence, the theoretical particle density in the film, m_p, as given by Eq. 28, has to be corrected for the actual particle diffusion coefficient. The impact of the several forces acting on a particle, i.e. diffusion and gravitation, was shown to depend mainly on the particle diameter (Fig. 59). Additionally, Vereecken et al. [51, 114] indicated that the gravitational force becomes important at a particle size of 300 nm. An increase in the particle size leads to a gradual reduction of the particle diffusion coefficient resulting in a decreasing impact of the diffusion. Hence, due to the large alumina agglomerates possibly present in the plating electrolyte, the particle transport can not be described solely by a diffusion process because an increase in the particle size leads to an increasing impact of the gravitational force. Moreover, the gravitational force, represented by the parameter A_G (Eq. 26), has to be considered in the calculation of the theoretical particle density in the film. Combining Eq. 26 and 28 gives an expression for m_p taking into account particle transport by diffusion and gravitation

$$m_p = \frac{zF}{V_{m,M}\,i}\cdot\left(\frac{0.9\,\upsilon^{1/2} D_p^{2/3}}{\nu^{1/6} H^{0.09} d^{0.41}} + \frac{(\rho_p - \rho_f)g r^2}{9\eta}\right)\cdot n_b \qquad (29)$$

4.2 Comparison with Experimental Results

The experimental data of the ECD of Ni-Al$_2$O$_3$ composites with the IJE system was extensively described in section 3.4. ECD experiments were done at three electrolyte flow rates (1, 2.5, and 6.5 l min^{-1}), four current densities (5, 10, 15, and 20 A dm^{-2}) and various particle loadings of the electrolyte (0-120 g l^{-1}).

Figure 60 shows the particle density in the composite film, m_p, as a function of the particle density in the electrolyte, n_b, for various IJE flow rates and plating current densities. The solid lines refer to the theoretical predictions calculated according to Eq. 28 (p. 90), whereas the open squares are related to the experimental data (section 3.4).

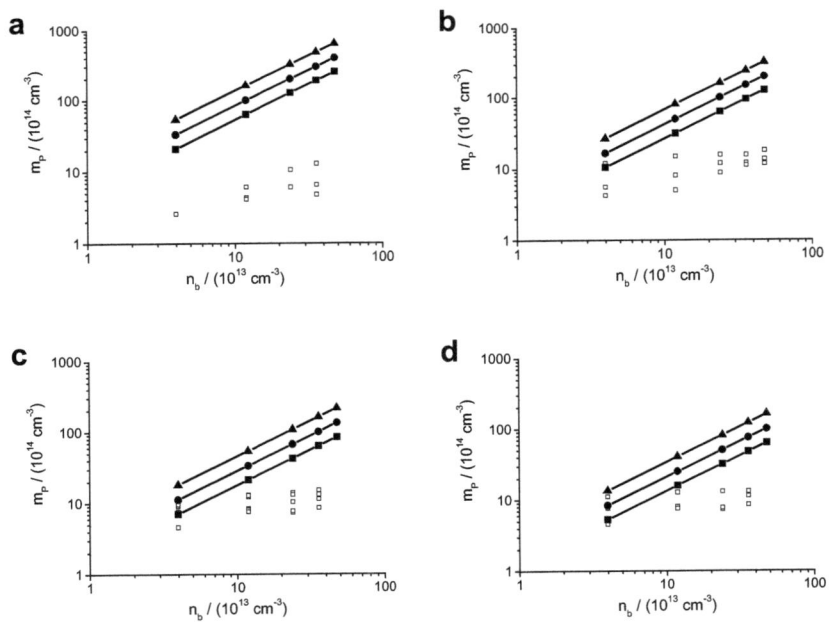

Figure 60 The number of nanoparticles per unit volume of nickel in the film, m_p, vs. particle number density in the solution, n_b, for flow rates of 1-6.5 l min^{-1}. The solid lines are the calculated particle number densities in the film under diffusion-limiting conditions (IJE flow rate: (■) 1 l min^{-1}, (●) 2.5 l min^{-1}, (▲) 6.5 l min^{-1}) and the open squares refer to the experimental results. (a) 5 A dm^{-2}, (b) 10 A dm^{-2}, (c) 15 A dm^{-2}, (d) 20 A dm^{-2}.

The experimental results (Fig. 60, symbols) are significantly smaller compared to the theoretical predictions (Fig. 60, solid lines). However, the overall results emphasize that the theoretical and experimental data converge with increasing plating current density (Fig. 60). The probability of particle incorporation significantly depends on the metal deposition rate and hence on the applied current density (section 1.4.4). When the metal growth rate is slow compared to the residence time of particles, the particles can diffuse from the electrode before becoming incorporated into the composite film. On the other hand, if the metal growth rate is fast in comparison to the residence time all particles adsorbed on the electrode become incorporated in the film. Finally, the probability of particle incorporation is a function of the metal growth rate and hence of the applied current density. The probability is equal to one when the metal growth rate is sufficiently high. Furthermore, the variations observed in Fig. 60 might be due to the fact that the actual particle velocity in the electrolyte is lower, *i.e.* the particle flux to the electrode is lower. It has to be emphasized that all the calculations so far depend on the primary particle diameter as given by the producer. However, the diameter of the alumina particles will be different depending on the surrounding media. The particle size was determined in the nickel sulfamate electrolyte by means of photon correlation spectroscopy (section 2.2.2). Of course, the alumina nanoparticles tend to aggregate in the nickel bath probably due to the high ionic strength (section 1.2.2). Assuming a normal distribution these aggregates are characterized by an average size of about 350 nm and a full width at half maximum of 125 nm (Fig. 61). Hence, it is assumed that the plating electrolyte contains alumina particles with a diameter ranging from 200 to 500 nm.

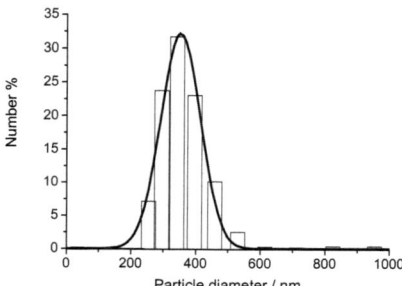

Figure 61   Particle size distribution of the alumina nanoparticles in the acidic nickel sulfamate electrolyte determined by photon correlation spectroscopy.

As discussed in section 4.1, the two transport parameters A_D and A_G, describing the particle transport to the electrode by means of diffusion and gravitation, exhibit a noticeable dependence on the particle diameter (Eqs. 25 and 26, p. 89). While the parameter A_D predominates in the range of small particles, the impact of the parameter A_G increases considerably with increasing diameter (Fig. 59, p. 89). Referring to the particle size analysis in the nickel electrolyte (Fig. 61), both transport mechanism have to be considered for the theoretical predictions of particle incorporation (Eq. 29, p. 91).

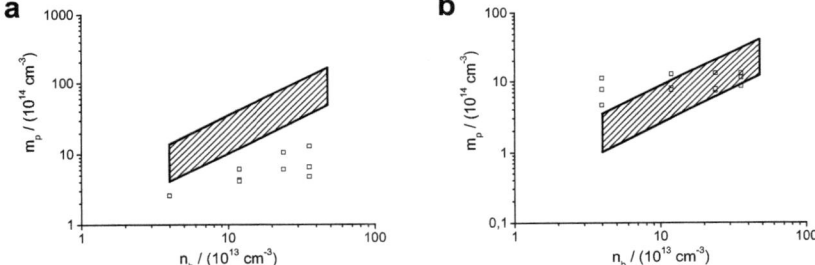

Figure 62 The number of nanoparticles per unit volume of nickel in the film, m_p, vs. particle number density in the solution, n_b, for IJE flow rates of 1-6.5 l min^{-1} and a current density of (a) 5 A dm^{-2} or (b) 20 A dm^{-2}, respectively. The hatched areas refer to the theoretical particle number densities in the film taking into account the actual particle size distribution in the electrolyte (Fig. 61) and the open squares refer to the experimental results.

According to Eq. 29 (p. 91), the theoretical prediction of particle incorporation is given by the hatched areas in Fig. 62 for various electrolyte flow rates and particle sizes between 200 and 500 nm. As can be seen in Fig. 62a, the experimental data is obviously below the diffusion-limiting case. This observation might be due to the fact that the nanoparticle concentration at the electrode surface is non-zero, particularly in the range of high particle contents of the electrolyte [114]. In accordance with this assumption, the theoretical predictions and the experimental results converge at higher plating current densities (Fig. 62b). Regardless of the particular plating current density, the overall discrepancies between experimental and theoretical data observed in Fig. 62 in the range of low as well as high particle loadings of the electrolyte are most probably due to particle interactions in solution or differences in the particle adsorption onto the electrode. Both processes are mainly governed by the interplay of either particle-particle or particle-electrode short range forces, respectively.

4.3 Summary

A kinetic model for ECD with an unsubmerged IJE system was developed to predict the amount of alumina incorporation in a nickel matrix plated from an acidic sulfamate electrolyte (section 4.1). The particle flux to the electrode was derived from a detailed analysis of the total force acting on a rigid spherical particle in an unbounded fluid in front of the electrode. The forces considered are convectional diffusion, gravity and buoyancy.

The comparison of the theoretical predictions with the experimental results (Figs. 60, p. 91 and 62, p. 93) proved a predominant particle transport to the electrode by means of convectional diffusion. However, it has to be emphasized that the primary particle size of the alumina nanopowder as given by the producer can not be utilized for the prediction of the particle content of the plated composite films. In fact, the actual particle size in the nickel electrolyte has to be determined. Nevertheless, the experimental incorporation results are still obviously smaller compared to the theoretical predictions in the range of small current densities and high particle contents in the electrolyte (Fig. 62, p. 93). Hence, the surface concentration of the nanoparticles on the electrode might not be zero indicating that either the particle flux might not be under diffusion control or that the metal plating rate, as defined by the applied current density, is not sufficiently high compared to the residence time of particles on the electrode surface.

The process of ECD is assumed to proceed in several stages as described in section 1.3. Apart from the particle transport to the electrode, the probability of particle incorporation is affected by the particle adsorption onto the electrode. Therefore, the short-range forces acting between particle and electrode are of prime importance. Colloidal probe microscopy experiments are particularly promising to gain a better insight in the impact of these short range forces on the ECD process [135, 212].

5 Conclusions and Suggested Future Work

The objectives of this research were to (1) characterize the nanoparticle properties (section 3.1), (2) systematically investigate the effects of certain process parameters on the ECD of nano-sized metal or metal oxide particles in a nickel matrix (sections 3.3-3.5), (3) examine how the particle incorporation affects the structure and properties of the electrodeposited nickel matrix (section 3.6), and (4) use the knowledge gained from the codeposition experiments to model the ECD process (section 4).

The key results obtained in this work, which were discussed in detail in the previous sections (3 and 4) will now be summarized. Suggestions for (possible) future work are given at the end of this section.

5.1 Conclusions for the Particle Characterization

The hydrodynamic diameter, surface charge and sedimentation behavior of the nanoparticles, *i.e.* Al_2O_3 (section 3.1.1), TiO_2 (section 3.1.2), Co (section 3.1.3), Fe_3O_4 (section 3.1.4), in a nickel electrolyte were characterized as a function of the composition, the ionic strength and the pH of the electrolyte. The nanoparticles feature a spherical or cubic shape with a diameter of 10-60 nm. Measurements of the zeta potential showed that the particles are positively charged in the acidic sulfamate bath and negatively charged in the alkaline pyrophosphate bath (Fig. 16, p. 42, Fig. 18, p. 43 and Fig. 25, p. 48). The hydrodynamic diameter of the particles in the aqueous electrolytes is directly related to their surface charge. An increase in the hydrodynamic diameter was found with decreasing absolute value of the zeta potential (Fig. 25, p. 48). Regardless of the type of particles, the colloidal stability of the nanoparticle within the nickel electrolytes was distinctly affected by the dispersant conditions such as pH and ionic strength (Fig. 20, p. 45).

5.2 Conclusions for the Electrocodeposition of Nickel Composite Films

The ECD was carried out using various current modulations (*i.e.* DC, PP and PRP) and electrode configurations (*i.e.* PPE and IJE). Additionally, the influence of certain deposition parameters such as pH, composition and particle content of the bath as well as current density, on the amount and the distribution of incorporated nanoparticles has been analyzed. The effects of the deposition parameters on the ECD process were investigated by evaluating the particle content using electrogravimetric analysis, EDX and GD-OES. A maximum incorporation of 12 vol-% of 50 nm Al_2O_3 in a nickel matrix was achieved using an IJE system (section 3.4), while PP and PRP resulted in composites with particle contents up to 11 vol-% of 13 nm Al_2O_3 particles (section 3.3.2). In comparison, DC depositions with the PPE system yielded composites with a maximum particle content either of ~3.6 vol-% of 13 nm Al_2O_3 or ~10.4 vol-% of 21 nm TiO_2 (section 3.3.1).

The effect of the composition and pH of the electrolyte on the ECD of Ni-Al$_2$O$_3$ and Ni-TiO$_2$ composites (section 3.3) was studied using the PPE configuration and DC deposition. The volume fraction of alumina (Fig. 28, p. 51) as well as titania (Fig. 29, p. 52) in the layer showed a tendency to increase with increasing particle content of the electrolyte and decreasing current density. It has been shown that negatively charged particles (both Al$_2$O$_3$ as well as TiO$_2$) co-deposit preferentially within the nickel matrix. This finding, which is a little bit counterintuitive at the first glance, was explained with an electrostatic model which takes into consideration the presence of ionic clouds around the particle and on the electrode, respectively.

Furthermore, the effect of PP and PRP on the ECD of Ni-Al$_2$O$_3$ composites was investigated using a PPE configuration. PP depositions were performed using rectangular pulses with pulse frequencies ranging from 1.7 to 8 Hz, pulse duty cycles of 11-67 % and a cathodic peak current density of either 5 or 10 A dm^{-2}. The particle content of the composite films increased with decreasing duty cycle and increasing pulse frequency (Fig. 30, p. 54). In PRP depositions, the incorporation of 13 nm Al$_2$O$_3$ particles was obviously improved due to longer cathodic pulse times and lower anodic peak current densities (Fig. 31, p. 55). High amounts of particle incorporation were found at low values of the average current density. The amount of co-deposited alumina ranged from 2 to 11 vol-% Al$_2$O$_3$ depending on the current modulation and the plating parameters (Fig. 31, p. 55).

An impinging jet electrode was used to study the influence of hydrodynamics on the ECD of Ni-Al$_2$O$_3$ composites from an acidic sulfamate electrolyte. The process parameters studied with the IJE system were flow rate, current density and particle loading of the electrolyte. Maximum particle incorporation was found at flow rates of 2-3 l min^{-1} (Fig. 33, p. 58), at a current density of 10-15 A dm^{-2} (Fig. 35, p. 59), and at high particle loadings of the electrolyte (Fig. 34, p. 58).

The effect of an external magnetic field on the codeposition of Co or Fe$_3$O$_4$ nanoparticles with nickel was studied (section 3.5). It has been shown, that the ECD of these nanoparticles in a Ni matrix is a versatile technique for the preparation of magnetic thin films. The amount of particle incorporation ranged between 20 to 41 vol-% Co (Table 4, p. 62) and 0 to 4 vol-% Fe$_3$O$_4$ (Fig. 39, p. 63). In general, decreasing current density and increasing particle content of the electrolyte was found to enhance the particle codeposition. The effect of a magnetic field on the particle incorporation has been shown using two different orientations of the B field. In a parallel field the particle codeposition decreased, whereas in a perpendicular field the particle content of the composite films showed a tendency to increase (Table 4, p. 62). The improved particle codeposition in the perpendicular orientation is most probably due to the magnetophoretic force, which leads to an increase in the particle concentration at the electrode surface.

5.3 Conclusions for the Structure and Properties of Nickel Composite Films

The structure (sections 3.6.1 and 3.6.2) as well as the mechanical (section 3.6.4) and magnetic properties (section 3.6.5) of the composite films have been investigated as a function of the particle content.

Surface morphology (Figs. 40 and 41, p. 65-66) and microstructure (Figs. 42-44, p. 67-68) of the nickel matrix were significantly altered due to the codeposition of nanoparticles. The pure nickel films from the sulfamate bath showed a field-oriented nickel growth (Fig. 42, p. 67). Due to the nanoparticle incorporation the microstructure of the nickel matrix changed from columnar to granular (Fig. 43, p. 68). In the case of the pyrophosphate bath the pure nickel films as well as the nanocomposites exhibited a granular nickel growth (Fig. 44, p. 68).

A strong (100) preferred orientation was found for the pure nickel film deposited from the sulfamate bath, which has been reduced due to a change in the plating conditions and the particle incorporation (Fig. 45, p. 69). The nanosized particles were found to inhibit the nickel growth and thus lead to a smaller crystallite size (Table 6, p. 70) combined with a distinct loss of texture (Table 5, p. 70). The pyrophosphate bath promoted a Ni crystal growth in direction of the (111) plane (Fig. 45, p. 69). Both, increasing current density and the particle incorporation became manifested in a loss of texture (Table 5, p. 70).

EBSD mapping indicated that the inclusion of alumina nanoparticles preferentially takes place in the grain boundary region, where the particles somehow terminate the nickel growth (Fig. 47, p. 73). High-resolution TEM imaging of nanocomposites proved the incorporation of alumina nanoparticles in the nickel matrix without any voids (Figs. 52 and 53, p. 78). Particle incorporation appeared in the form of single particles as well as chains of particles (Fig. 51, p. 77). Moreover, the observation of lattice planes of the polycrystalline nickel matrix gives a relatively clear picture about the interface between particle and matrix (Fig. 53, p. 78).

The mechanical properties of the composite films were found to be dominated by the crystallite size (Table 6, p. 70) and structure of the nickel matrix (Fig. 42-44, p. 67-68). In the case of both nickel baths, the Vickers microhardness showed a tendency to increase with the amount of particle incorporation and with decreasing current density (Fig. 54, p. 79). According to the variation of the microhardness, the abrasion resistance under unlubricated conditions increased with decreasing current density (Fig. 55, p. 81).

The embedding of magnetic nanoparticles was found to affect the magnetic response, *i.e.* saturation magnetization and coercivity, of the Ni matrix considerably. The magnetic hardness of the Ni films was strongly increased by the incorporation of Co nanoparticles (Table 8, p. 82). Moreover, the presence of a static magnetic field during the ECD influences the amount of nanoparticle incorporation (Table 4, p. 62) which substantially affects the magnetic hardness of the films (Table 8, p. 82).

5.4 Conclusions for the Modeling of the Electrocodeposition

The experimental results of the ECD of Ni-Al$_2$O$_3$ composites using the impinging jet electrode system (section 3.4) have been described by a kinetic model based on a detailed analysis of the particle flux to the electrode (section 4.1). The particle flux was derived from an analysis of the total force acting on a rigid spherical particle in an unbounded fluid in front of the electrode of the IJE system (Fig. 57, p. 84). The forces considered are the convectional diffusion, gravity and buoyancy. The influence of gravitational and buoyancy force were found to be negligible for 50 nm particles, but they were identified to be important for bigger agglomerates in the range of 300-500 nm (Fig. 59, p. 89). Comparison of the theoretical predictions values with the experimental results (Fig. 62, p. 93) revealed discrepancies in the range of low as well as high particle loadings of the electrolyte which can be explained with particle interactions in solution or differences in the particle adsorption behavior. Both processes are mainly governed by the interplay of either particle-particle or particle-electrode short range forces, respectively. Moreover, the surface concentration of nanoparticles on the electrode might not be zero indicating that the particle flux might not be under diffusion control.

5.5 Suggested Future Work

It has been shown in the present work, that both jet plating and pulse plating improved the incorporation of alumina nanoparticles in a nickel matrix. However, different types of alumina nanopowder have been used for the IJE and PPE experiments. Further experiments and analysis with the IJE system are needed using the 13 nm alumina nanoparticles. Additionally, a combination of both deposition techniques, *i.e.* jet and pulse plating, seems promising in terms of an improvement of the amount and distribution of particles in the composite film.

The kinetic model on ECD developed in this research should be improved or expanded by further ECD experiments using the IJE system with varying particle types, sizes and surface properties. Apart from the particle transport, the probability of particle incorporation is affected by the particle adsorption onto the electrode. Therefore the short-range forces acting between particle and electrode are of prime importance. Colloidal probe microscopy experiments can be utilized to gain a better insight in these short range forces.

The recent particle size and the dispersion stability measurements have shown that one of the most critical aspects of using nanoparticles in plating baths with high ionic strengths is the difficulty to keep them in suspension and to prevent the agglomeration. A surface treatment of the particles by grafting of polymer brushes can enable a combined electrostatic and steric stabilization of the colloidal dispersion. A study utilizing nanoparticles of the same chemical composition and with the same mean diameter, but with different surface properties could help to further the understanding of the mechanism by which particles are entrapped during the ECD process. Furthermore, additives for the purpose of de-agglomeration of nanoparticles in electrolytes with high ionic strengths should be explored. It is conceivable that smaller particle sizes in the electrolyte also have significant effects on grain size and structure of the metal matrix, which eventually affects the properties of the composite films. Smaller particles would create more sites at which grain boundary pinning would occur.

Another influential surface property of the nanoparticles is their hydrophilicity, *i.e.* the tendency of getting solvated by water. Due to the formation of a hydration layer around the particles, they remain separated from the electrode by a small gap which substantially affects the particle embedding by the growing metal matrix. The degree of hydration, *i.e.* the thickness of the hydration layer, can be influenced by a proper selection of the operating conditions such as particle characteristics and electrolyte composition. Aqueous electrolytes are most commonly used for galvanic applications. A partial or complete substitution of the water by organic solvents (*e.g.* methanol, ethanol, etc.) can affect the state of hydration and therefore influence the interaction between particle and electrode. Finally, the particle incorporation behavior might change which can substantially alter the film properties.

Due to the nanoparticle incorporation, the structure and properties of the metal matrix was found to change considerably. With reference to an industrial application a detailed characterization of the chemical, electrical, and mechanical film properties is required. Particularly, the effect of functional nanoparticles, *e.g.* microcapsules, on the strengthening and corrosion mechanism of the metal matrix has to be elucidated in view of an optimization of the film properties.

For the purpose of scaling-up or industrialization of the ECD process, the use of experimental setups with a practical orientation combined with bigger and complexly shaped substrates is mandatory. Additionally, a plan of procedures for electrolyte monitoring, with special regard to a possible aging of the electrolyte dispersions, should be developed.

6 References

[1] J.-P. Celis, J.R. Roos, C. Buelens, *J. Electrochem. Soc.* 134 (1987) 1402.
[2] J.L. Stojak, J. Fransaer, J.B. Talbot, *Review of Electrocodeposition*, in: R.C. Alkire, Kolb, D.M. (Eds.), *Adv. Electrochem. Sci. Eng.*, Wiley-VCH Verlag, Weinheim, 2002.
[3] C.T.J. Low, R.G.A. Wills, F.C. Walsh, *Surf. Coat. Technol.* 201 (2006) 371.
[4] N.R.d. Tacconi, H. Wenren, K. Rajeshwar, *J. Electrochem. Soc.* 144 (1997) 3159.
[5] V. Medeliene, V. Stankevic, G. Bikulcius, *Surf. Coat. Technol.* 168 (2003) 161.
[6] M. Zhou, N.R.d. Tacconi, K. Rajeshwar, *J. Electroanal. Chem.* 421 (1997) 111.
[7] N.R.d. Tacconi, C.A. Boyles, K. Rajeshwar, *Langmuir* 16 (2000) 5665
[8] D. Thiemig, C. Kubeil, C.P. Gräf, A. Bund, *Thin Solid Films (submitted)*.
[9] A. Roldan, E. Gómez, S. Pané, E. Vallés, *J. Appl. Electrochem.* 37 (2007) 575.
[10] S. Veprek, A.S. Argon, *Surf. Coat. Technol.* 146-147 (2001) 175.
[11] A.F. Zimmerman, G. Palumbo, K.T. Aust, U. Erb, *Mater. Sci. Eng.* A328 (2002) 137.
[12] Y.-J. Xue, D. Zhu, F. Zhao, *J. Mater. Sci.* 39 (2004) 4063.
[13] S. Tao, D.Y. Li, *Nanotechnol.* 17 (2006) 65.
[14] E.W. Brooman, *Galvanotechnik* (2006) 58.
[15] E. Brooman, *Plat. Surf. Finish.* 94 (2007) 38.
[16] A. Bund, D. Thiemig, *Surf. Coat. Technol.* 201 (2007) 7092.
[17] A. Bund, D. Thiemig, *J. Appl. Electrochem.* 37 (2007) 345.
[18] B. Bozzini, G. Giovannelli, P.L. Cavallotti, *J. Microscop.* 185 (1997) 283.
[19] L. Wang, Y. Gao, Q. Xue, H. Liu, T. Xu, *Mater. Sci. Eng.* A390 (2005) 313.
[20] P. Cojocaru, A. Vicenzo, P.L. Cavallotti, *J. Solid State Electrochem.* 9 (2005) 850–858.
[21] V. Terzieva, J. Fransaer, J.-P. Celis, *J. Electrochem. Soc.* 147 (2000) 198.
[22] Y. Zhang, Y. Fan, X. Yang, Z. Chen, J. Zhang, *Plat. Surf. Finish.* (2004) 39.
[23] G. Vidrich, J.-F. Castagnet, H. Ferkel, *J. Electrochem. Soc.* 152 (2005) C294.
[24] X. Bin-shi, W. Hai-dou, D. Shi-yun, J. Bin, T. Wei-yi, *Electrochem. Comm.* 7 (2005) 572.
[25] P.-A. Gay, P. Berçot, J. Pagetti, *Surf. Coat. Technol.* 140 (2001) 147.
[26] W. Wang, F.Y. Hou, H. Wang, H.T. Guo, *Scripta Mater.* 53 (2005) 613.
[27] F. Hou, W. Wang, H. Guo, *Appl. Surf. Sci.* 252 (2006) 3812.
[28] S.W. Watson, *J. Electrochem. Soc.* 140 (1993) 2235.
[29] P. Nowak, R.P. Socha, M. Kaisheva, J. Fransaer, J.-P. Celis, Z. Stoinov, *J. Appl. Electrochem.* 30 (2000) 429.
[30] A.F. Zimmerman, D.G. Clark, K.T. Aust, U. Erb, *Mater. Lett.* 52 (2002) 85.
[31] F. Hu, K.C. Chan, *Appl. Surf. Sci.* 233 (2004) 163.
[32] M. Kaisheva, J. Fransaer, *J. Electrochem. Soc.* 151 (2004) C89.
[33] A. Hovestad, L.J.J. Janssen, *J. Appl. Electrochem.* 25 (1995) 519.

[34] M. Musiani, *Electrochim. Acta* 45 (2000) 3397.
[35] J.-P. Celis, J. Fransaer, *Galvanotechnik* 88 (1997) 2229.
[36] T.-S. Chin, *J. Magn. Magn. Mater* 209 (2000) 75.
[37] D. Niarchos, *Sens. Actuators, A* A109 (2003) 166.
[38] M.R.J. Gibbs, E.W. Hill, P.J. Wright, *J. Phys. D: Appl. Phys.* 37 (2004) R237.
[39] W. Wang, Z. Yao, J.C. Chen, J. Fang, *J. Micromech. Microeng.* 14 (2004) 1321.
[40] M.R.J. Gibbs, *J. Magn. Magn. Mater* 290-291 (2005) 1298.
[41] E. Gómez, S. Pané, E. Vallés, *Electrochem. Comm.* 7 (2005) 1225.
[42] B.Y. Yoo, S.C. Hernandez, D.-Y. Park, N.V. Myung, *Electrochim. Acta* 51 (2006) 6346.
[43] M. Onoda, K. Shimizu, T. Tsuchiya, T. Watanabe, *J. Magn. Magn. Mater* 126 (1993) 595.
[44] T. Osaka, M. Takai, K. Hayashi, K. Ohashi, M. Satto, K. Yamada, *Nature* 392 (1998) 796.
[45] M. Takai, K. Hayashi, M. Aoyagi, T. Osaka, *J. Electrochem. Soc.* 144 (1997) L203.
[46] J.W. Judy, R.S. Muller, *Sens. Actuators, A* A53 (1996) 392.
[47] N.V. Myung, D.-Y. Park, B.-Y. Yoo, P.T.A. Sumodjo, *J. Magn. Magn. Mater* 265 (2003) 189.
[48] Y. Su, H. Wang, G. Ding, F. Cui, W. Zhang, W. Chen, *IEEE Trans. Magn.* 41 (2005) 4380.
[49] N. Guglielmi, *J. Electrochem. Soc.* 119 (1972) 1009.
[50] K. Helle, F. Walsh, *Trans. Inst. Met. Finish.* 75 (1997) 53.
[51] P.M. Vereecken, I. Shao, P.C. Searson, *J. Electrochem. Soc.* 147 (2000) 2572.
[52] J. Fransaer, J.-P. Celis, *Galvanotechnik* 92 (2001) 1544.
[53] S.-C. Wang, W.-C.J. Wei, *Mater. Chem. Phys.* 78 (2003) 574.
[54] P. Xiong-Skiba, D. Engelhaupt, R. Hulguin, B. Ramsey, *J. Electrochem. Soc.* 152 (2005) C571.
[55] J. Fransaer, J.-P. Celis, J.R. Roos, *J. Electrochem. Soc.* 139 (1992) 413.
[56] D.L. Wang, J. Li, C.S. Dai, X.G. Hu, *J. Appl. Electrochem.* 29 (1999) 437.
[57] F. Wünsche, A. Bund, W. Plieth, *J. Solid State Electrochem.* 3 (2004) 209.
[58] J.L. Stojak, J.B. Talbot, *J. Electrochem. Soc.* 146 (1999) 4504.
[59] A. Hovestad, R.J.C.H.L. Heesen, L.J.J. Janssen, *J. Appl. Electrochem.* 29 (1999) 331.
[60] L. Stappers, J. Fransaer, *J. Electrochem. Soc.* 152 (2005) C392.
[61] C.M.A. Brett, A.M.O. Brett, *Electrochemistry Principles, Methods, and Applications*, Oxford University Press, Oxford, New York, Tokyo, 1993.
[62] R.C. Alkire, T.J. Chen, *J. Electrochem. Soc.* 129 (1982) 2424.
[63] C. Karakus, D.T. Chin, *J. Electrochem. Soc.* 141 (1994) 691.
[64] M. Okumiya, H. Takeuchi, Y. Tsunekawa, *J. Jpn. Inst. Met.* 59 (1995) 640.
[65] H. Takeuchi, Y. Tsunekawa, M. Okumiya, *Mater. Trans.* 38 (1997) 43.

[66] H. Takeuchi, S. Tamura, Y. Tsunekawa, M. Okumiya, *Surf. Eng.* 20 (2004) 25.
[67] I. Garcia, A. Conde, G. Langelaan, J. Fransaer, J.-P. Celis, *Corr. Sci.* 45 (2003) 1173.
[68] E.B. Budevski, G.T. Staikov, W.J. Lorenz, *Electrochemical Phase Formation and Growth. An Introduction to the Initial Stages of Metal Deposition*, Wiley-VCH, Weinheim, 1996.
[69] J.C. Puippe, *Pulse-Plating*, Leuze Verlag, Bad Saulgau, 1986.
[70] C. Kollia, Z. Loizos, N. Spyrellis, *Surf. Coat. Technol.* 45 (1991) 155.
[71] P.T. Tang, T. Watanabe, J.E.T. Andersen, G. Bech-Nielsen, *J. Appl. Electrochem.* 25 (1995) 347.
[72] R. Mishra, R. Balasubramaniam, *Corr. Sci.* 46 (2004) 3019.
[73] C.Y. Dai, Y. Pan, S. Jiang, Y.C. Zhou, *Surf. Rev. Lett.* 11 (2004) 433.
[74] A.B. Vidrine, E.J. Podlaha, *J. Appl. Electrochem.* 31 (2001) 461.
[75] M.E. Bahrololoom, R. Sani, *Surf. Coat. Technol.* (2005) 154.
[76] E.A. Pavlatou, M. Stroumbouli, P. Gyftou, N. Spyrellis, *J. Appl. Electrochem.* 36 (2006) 385.
[77] E.J. Podlaha, D. Landolt, *J. Electrochem. Soc.* 144 (1997) L200.
[78] E.J. Podlaha, *Nano Lett.* 1 (2001) 413.
[79] B. Müller, H. Ferkel, *Nanostruct. Mater.* 10 (1998) 1285.
[80] J. Steinbach, H. Ferkel, *Scripta Mater.* 44 (2001) 1813.
[81] A.M.J. Kariapper, J. Foster, *Trans. Inst. Met. Finish.* 52 (1974) 87.
[82] D. Baudrand, *Met. Fin.* 94 (1996) 15.
[83] M. Schlesinger, M. Paunovic, *Modern Electroplating*, John Wiley & Sons, New York, 2000.
[84] C.S. Lin, C.Y. Lee, C.F. Chang, C.H. Chang, *Surf. Coat. Technol.* 200 (2006) 3690.
[85] J.B. Talbot, *Plat. Surf. Finish.* 91 (2004) 60.
[86] L. Shi, C. Sun, P. Gao, F. Zhou, W. Liu, *Appl. Surf. Sci.* 252 (2006) 3812.
[87] N.S. Qu, D. Zhu, K.C. Chan, *Scripta Mater.* 54 (2006) 1421.
[88] S. Steinhäuser, B. Wielage, *Surf. Eng.* 13 (1997) 289.
[89] G.C. Fink, J.D. Prince, *Trans. Am. Electrochem. Soc.* 54 (1928) 315
[90] E.C. Kedward, K.W. Wright, *Plat. Surf. Finish.* 65 (1978) 38.
[91] V.P. Greco, *Plat. Surf. Finish.* 76 (1989) 62.
[92] V.P. Greco, *Plat. Surf. Finish.* 76 (1989) 68.
[93] H. Abi-Akar, C. Riley, *Chem. Mater.* 8 (1996) 2601
[94] J.R. Groza, J.C. Gibeling, *Mater. Sci. Eng. A* 171 (1993) 115
[95] J.L. Stojak, *An Investigation of Electrocodeposition using a Rotating Cylinder Electrode*, PhD thesis, University of California, San Diego, 1997.
[96] G. Cârâc, A. Bund, D. Thiemig, *Surf. Coat. Technol.* 202 (2007) 403.
[97] P.R. Webb, N.L. Robertson, *J. Electrochem. Soc.* 141 (1994) 669.

[98] J.P. Celis, H. Kelchtermans, J.R. Roos, *Trans. Inst. Met. Fin.* 56 (1978) 41.
[99] Z. Serhal, J. Morvan, M. Rezrazi, P. Berçot, *Surf. Coat. Technol.* 140 (2001) 166.
[100] J.L. Valdes, *Deposition of Colloidal Particles in Electrochemical Systems*, PhD thesis, Columbia University, New York, 1987.
[101] R.J. Hunter, *Foundations of Colloid Science*, 2nd edition, Oxford University Press, Oxford, New York, 2001.
[102] A.J. Bard, L.R. Faulkner, *Electrochemical methods. Fundamentals and applications*, 2nd Edition, John Wiley and Sons, New York, 2001.
[103] G. Gouy, *J. Physique* 9 (1910) 457.
[104] D.L. Chapman, *Phil. Mag.* 25 (1913) 475.
[105] *Zetasizer Nano Series User Manual*, Issue 2.1, Malvern Instruments Ltd., Worcestershire, United Kingdom, July 2004.
[106] D. Myers, *Surfaces, Interfaces and Colloids - Principles and Applications*, 2nd edition, Wiley-VCH, New York, 1999.
[107] G.A. Parks, *Chem. Rev.* 65 (1965) 177.
[108] C.R. Evankoa, R.F. Delisioa, D.A. Dzombaka, J. J. W. Novak, *Colloids Surf., A* 125 (1997) 95.
[109] B.V. Deryagin, L. Landau, *Acta Physiochim. U.R.S.S.* 14 (1941) 633.
[110] E.J.W. Verwey, J.T.G. Overbeek, *Theory of the stability of lyophobic colloids*, Elsevier Publ. Comp., Amsterdam, New York, 1948.
[111] S. Steinhäuser, *Galvanotechnik* 92 (2001) 940.
[112] J.P. Celis, J.R. Roos, *J. Electrochem. Soc.* 124 (1977) 1508.
[113] J.W. Graydon, D.W. Kirk, *J. Electrochem. Soc.* 137 (1990) 2061.
[114] I. Shao, P.M. Vereecken, R.C. Cammarata, P.C. Searson, *J. Electrochem. Soc.* 149 (2002) C610.
[115] J. Lee, J.B. Talbot, *J. Electrochem. Soc.* 154 (2007) D70.
[116] A. Bund, D. Thiemig, *ECS Trans.* 3 (2006) 85.
[117] D. Thiemig, R. Lange, A. Bund, *Electrochim. Acta* 52 (2007) 7362.
[118] M. Eisenberg, C.W. Tobias, C.R. Wilke, *J. Electrochem. Soc.* 101 (1954) 306.
[119] J.S. Newman, K.E. Thomas-Alyea, *Electrochemical Systems*, 3rd edition, John Wiley & Sons Inc., Hoboken, New Jersey, 2004.
[120] Q. Li, J.D.A. Walker, *AIChE J.* 42 (1996) 391.
[121] D.-T. Chin, K.-L. Hsueh, *Electrochim. Acta* 31 (1986) 561.
[122] S.J. Osborne, *Electrocodeposition of Nanoparticle Composite Films Using an Impinging Jet Electrode*, PhD thesis, University of California, San Diego, 2005.
[123] C. Buelens, J.P. Celis, J.R. Roos, *J. Appl. Electrochem.* 13 (1983) 541.
[124] R. Peipmann, J. Thomas, A. Bund, *Electrochim. Acta* 52 (2007) 5808.
[125] T.Z. Fahidy, *J. Electrochem. Soc.* 18 (1973) 607.
[126] R.A. Tacken, L.J.J. Janssen, *J. Appl. Electrochem.* 25 (1995) 1.

[127] A. Bund, S. Koehler, H.H. Kuehnlein, W. Plieth, *Electrochim. Acta* 49 (2003) 147.
[128] G. Hinds, F.E. Spada, J.M.D. Coey, T.R.N. Mhíocháin, M.E.G. Lyons, *J. Phys. Chem. B* 105 (2001) 9487.
[129] R.P. Socha, P. Nowak, K. Laajalehto, J. Väyrynen, *Colloids Surf., A* 235 (2004) 45.
[130] N. Kanani, *Electroplating - Basic Principles, Processes and Practice*, Elsevier, Oxford, 2004.
[131] W. Plieth, *Electrochemistry for Materials Science*, Elsevier, Amsterdam, 2007.
[132] W.H. Safranek, *The Properties of Electrodeposited Metals and Alloys: A Handbook*, American Elsevier Publishing Company, New York, 1974.
[133] F. Ebrahimi, G.R. Bourne, M.S. Kelly, T.E. Matthews, *Nanostruct. Mater.* 11 (1999) 343.
[134] L.W. Bruch, M.W. Cole, E. Zaremba, *Physical Adsorption: Froces and Phenomena*, Oxford University Press Inc., New York, 1997.
[135] C. Dedeloudis, J. Fransaer, *Langmuir* 20 (2004) 11030.
[136] S.W. Banovic, K. Barmak, A.R. Marder, *J. Mater. Sci.* 33 (1998) 639.
[137] W.S. Sweet, *Investigations to Electrocodepsition using an Impinging Jet Electrode*, Master thesis, University of California, San Diego, 2006.
[138] S.J. Osborne, W.S. Sweet, K.S. Vecchio, J.B. Talbot, *J. Electrochem. Soc.* 154 (2007) D394.
[139] S.J. Osborne, W.S. Sweet, J.B. Talbot, *Electroplating Nanocomposite Films with an Impinging Jet Electrode*, AESF Annual International Technical Conference, American Electroplaters and Surface Finishers Society, La Jolla, 2005, pp. 185.
[140] K. Barmak, S.W. Banovic, C.M. Petronis, D.F. Susan, A.R. Marder, *J. Microscop.* 185 (1997) 265.
[141] H. Czichos, T. Saito, L. Smith (Eds.), *Springer Handbook of Materials Measurement Methods*, 1, Springer Verlag, Berlin, 2006.
[142] R.S. Kottada, A.H. Chokshi, *Scripta Mater.* 53 (2005) 887.
[143] M. Ettlinger, *Highly Dispersed Metallic Oxides Produced by the AEROSIL® Process*, Degussa.
[144] A.F. Holleman, E. Wiberg, *Lehrbuch der Anorganischen Chemie*, 101. Auflage, Walter de Gruyter & Co., Berlin, 1995.
[145] *Buehler Ltd., Material Safety Data Sheet - Buehler Micropolish II*, Lake Bluff, USA, 15.03.2007.
[146] C.P. Gräf, R. Birringer, A. Michels, *Phys. Rev. B* 73 (2006) 212401.
[147] R. Massart, *IEEE Trans. Magn.* 17 (1981) 1247.
[148] C. Galindo-Gonzáles, J.d. Vicente, M.M. Ramos-Tejada, M.T. López-López, F. Gonzáles-Cabarello, J.D.G. Durán, *Langmuir* 21 (2005) 4410.
[149] Y. Sahoo, A. Goodarzi, M.T. Swihart, T.Y. Ohulchansky, N. Kaur, E.P. Furlani, P.N. Prasad, *J. Phys. Chem.* 109 (2005) 3879.

[150] A. Ditsch, P.E. Laibinis, D.I.C. Wang, T.A. Hatton, *Langmuir* 21 (2005) 6006.
[151] G. Gnanaprakash, S. Mahadevan, T. Jayakumar, P. Kalyanasundaram, J. Philip, B. Raj, *Mater. Chem. Phys.* 103 (2007) 168.
[152] A. Bund, H.H. Kuehnlein, *J. Phys. Chem.* 109 (2005) 19845.
[153] R. Schumacher, *Angew. Chem. Int. Ed.* 29 (1990) 329.
[154] A. Bund, *Die Quarzmikrowaage in Rheologie und Elektrochemie: Fortschritte in der Signalauswertung durch Netzwerkanalyse*, Dissertation, Saarland University, Saarbrücken, 1999.
[155] A. Bund, G. Schwitzgebel, *Electrochim. Acta* 45 (2000) 3703.
[156] H.H. Kühnlein, *Elektrochemische Legierungsabscheidung zur Herstellung von Cu_2ZnSnS_4 Dünnschichtsolarzellen*, Dissertation, Technische Universität Dresden, 2007.
[157] G. Sauerbrey, *Z. Phys.* 155 (1959) 206.
[158] T. Sobisch, D. Lerche, *Colloid Polym. Sci.* 278 (2000) 369.
[159] M. Birkholz, *Thin Film Analysis by X-Ray Scattering*, Wiley-VCH, Weinheim, 2006.
[160] A.J. Schwartz, M. Kumar, B.L. Adams (Eds.), *Electron Backscatter Diffraction in Materials Science*, Kluwer Academic / Plenum Publishers, New York, 2000.
[161] D. Brandon, W.D. Kaplan, *Microstructural Characterization of Materials*, John Wiley & Sons, Chichester, 1999.
[162] D. Grunenberg, D. Sommer, K.H. Koch, *Fresenius Z. Anal. Chem.* 319 (1984) 665.
[163] R. Payling, D. Jones, A. Bengston (Eds.), *Glow Discharge Optical Emission Spectrometry*, John Wiley & Sons, Chichester, 1997.
[164] H.P. Klug, L.E. Alexander, *X-Ray Diffraction Procedures*, 2. ed., John Wiley & Sons, New York, 1974.
[165] D.H. Jeong, F. Gonzalez, G. Palumbo, K.T. Aust, U. Erb, *Scripta Mater.* 44 (2001) 493.
[166] S. Chikazumi, *International series of monographs on physics: Physics of Ferromagnetism*, 2nd Edition, Clarendon Press, Oxford, 1997.
[167] P.C. Hidber, T.J. Graule, L.J. Gauckler, *J. Am. Ceram. Soc.* 79 (1996) 1857.
[168] M. Kosmulski, P. Dahlsten, P. Próchniak, J.B. Rosenholm, *Colloids Surf., A* 301 (2007) 425.
[169] S.H. Behrens, D.G. Grier, *J. Chem. Phys.* 115 (2001) 6716.
[170] V.F. Puntes, K.M. Krishnan, A.P. Alivisatos, *Science* 291 (2001) 2115.
[171] V.F. Puntes, K.M. Krishnan, P. Alivisatos, *Appl. Phys. Lett.* 78 (2001) 2187.
[172] J. Waddell, S. Inderhees, M.C. Aronson, S.B. Dierker, *J. Magn. Magn. Mater* 297 (2006) 54.
[173] S. Sun, C.B. Murray, *J. Appl. Phys.* 85 (1999) 4325.
[174] I. Nedkov, M. Ausloos (Eds.), *Nano-Crystalline and Thin Film Magnetic Oxides*, Kluwer Academic Publishers, Dordrecht, 1999.
[175] W. Yu, T. Zhang, J. Zhang, X. Qiao, L. Yang, Y. Liu, *Mater. Lett.* 60 (2006) 2298.

[176] L. Campanella, *J. Electroanal. Chem.* 28 (1970) 228.
[177] D. Thiemig, A. Bund, *Surf. Coat. Technol.* 202 (2008) 2976.
[178] D. Thiemig, A. Bund, J.B. Talbot, *J. Electrochem. Soc.* 154 (2007) D510.
[179] B.J. Hwang, C.S. Hwang, *J. Electrochem. Soc.* 140 (1993) 979.
[180] E. Tóth-Kádár, I. Bakonyi, L. Pogány, Á. Cziráki, *Surf. Coat. Technol.* 88 (1996) 57.
[181] D. Thiemig, A. Bund, J.B. Talbot, *Electrochim. Acta* (2008) accepted.
[182] F. Erler, C. Jakob, H. Romanus, L. Spiess, B. Wielage, T. Lampke, S. Steinhäuser, *Electrochim. Acta* 48 (2003) 3063.
[183] M. Stroumbouli, P. Gyftou, E.A. Pavlatou, N. Spyrellis, *Surf. Coat. Technol.* 195 (2005) 325.
[184] W.-H. Lee, S.-C. Tang, K.-C. Chung, *Surf. Coat. Technol.* 120-121 (1999) 607.
[185] C. Kollia, N. Spyrellis, J. Amblard, M. Froment, G. Maurin, *J. Appl. Electrochem.* 20 (1990) 1025.
[186] V. Medeliene, *Surf. Coat. Technol.* 154 (2002) 104.
[187] L. Du, B. Xu, S. Dong, H. Yang, Y. Wu, *Surf. Coat. Technol.* 192 (2005) 311.
[188] H. Matsushima, A. Bund, W. Plieth, S. Kikuchi, Y. Fukunaka, *Electrochim. Acta* 53 (2007) 161.
[189] H. Matsushima, A. Ispas, A. Bund, W. Plieth, Y. Fukunaka, *J. Solid State Elecrochem.* 11 (2007) 737.
[190] A. Bund, A. Ispas, *J. Electroanal. Chem.* 575 (2005) 221.
[191] A. Ispas, H. Matsushima, W. Plieth, A. Bund, *Electrochim. Acta* 52 (2007) 2785.
[192] C.S. Lin, P.C. Hsu, L. Chang, C.H. Chen, *J. Appl. Electrochem.* 31 (2001) 925.
[193] F. Ebrahimi, Z. Ahmed, *J. Appl. Electrochem.* 33 (2003) 733.
[194] Á. Cziráki, B. Fogarassy, I. Geröcs, E. Tóth-Kádár, I. Bakonyi, *J. Mater. Sci.* 29 (1994) 4771.
[195] L. Benea, P.L. Bonora, A. Borello, S. Martelli, F. Wenger, P. Ponthiaux, J. Galland, *Solid State Ionics* 151 (2002) 89.
[196] T. Lampke, B. Wielage, D. Dietrich, A. Leopold, *Appl. Surf. Sci.* 253 (2006) 2399.
[197] S.K. Panikkar, T.L.R. Char, *J. Electrochem. Soc.* 106 (1959) 495.
[198] E.O. Hall, *Proc. Phys. Soc. Lond.* B64 (1951) 747.
[199] N.J. Petch, *J. Iron Steel Inst.* 174 (1953) 25.
[200] I. Garcia, J. Fransaer, J.-P. Celis, *Surf. Coat. Technol.* 148 (2001) 171.
[201] W. Kleinekathöfer, C.J. Raub, *Surf. Technol.* 7 (1978) 23.
[202] W. Kleinekathöfer, C.J. Raub, E. Raub, *Metalloberfläche* 36 (1982) 411.
[203] M. Klingenberg, E.W. Brooman, T.A. Naguy, *Plat. Surf. Finish.* 92 (2005) 42.
[204] L. Chen, L. Wang, Z. Zeng, T. Xu, *Surf. Coat. Technol.* 201 (2006) 599.
[205] F. Cadieu, T. Cheung, L. Wickramasekara, N. Kamprath, *IEEE Trans. Magn.* 22 (1986) 752.
[206] L. Castaldi, M.R.J. Gibbs, H.A. Davies, *J. Appl. Phys.* 96 (2004) 5063.

[207] S.L. Tang, M.R.J. Gibbs, H.A. Davies, Z.W. Liu, S.C. Lane, N.E. Mateen, *J. Appl. Phys.* 101 (2007) 013910/1.

[208] S.L. Tang, M.R.J. Gibbs, H.A. Davies, Z.W. Liu, S.C. Lane, N.E. Mateen, Y.W. Du, *J. Appl. Phys.* 101 (2007) 09K501/1.

[209] T.-J. Chen, *Selective Jet Plating*, PhD Thesis, University of Illinois, Urbana-Champaign, 1981.

[210] C. Karakus, *Metal Thickness Distribution in Jet Plating*, Master Thesis, Clarkson University, 1992.

[211] *CRC Handbook of Chemistry and Physics on CD-ROM*, Version 2006, CRC Press LLC., 2006.

[212] L. Stappers, J. Fransaer, *J. Electrochem. Soc.* 154 (2007) D598.

Die VDM Verlagsservicegesellschaft sucht für wissenschaftliche Verlage abgeschlossene und herausragende

Dissertationen, Habilitationen, Diplomarbeiten, Master Theses, Magisterarbeiten usw.

für die kostenlose Publikation als Fachbuch.

Sie verfügen über eine Arbeit, die hohen inhaltlichen und formalen Ansprüchen genügt, und haben Interesse an einer honorarvergüteten Publikation?

Dann senden Sie bitte erste Informationen über sich und Ihre Arbeit per Email an *info@vdm-vsg.de*.

Sie erhalten kurzfristig unser Feedback!

VDM Verlagsservicegesellschaft mbH
Dudweiler Landstr. 99
D - 66123 Saarbrücken
www.vdm-vsg.de

Telefon +49 681 3720 174
Fax   +49 681 3720 1749

Die VDM Verlagsservicegesellschaft mbH vertritt

Printed by Books on Demand GmbH, Norderstedt / Germany